Key issues for mountain areas

Edited by Martin F. Price, Libor F. Jansky, and
Andrei A. Iastenia

United Nations
University Press

TOKYO · NEW YORK · PARIS

© United Nations University, 2004

The views expressed in this publication are those of the authors and do
not necessarily reflect the views of the United Nations University.

United Nations University Press
United Nations University, 53–70, Jingumae 5-chome,
Shibuya-ku, Tokyo, 150-8925, Japan
Tel: +81-3-3499-2811   Fax: +81-3-3406-7345
E-mail: sales@hq.unu.edu
general enquiries: press@hq.unu.edu
http://www.unu.edu

United Nations University Office at the United Nations, New York
2 United Nations Plaza, Room DC2-2062, New York,
NY 10017, USA
Tel: +1-212-963-6387   Fax: +1-212-371-9454
E-mail: unuona@ony.unu.edu

United Nations University Press is the publishing division of the United Nations
University.

Cover design by Mea Rhee

Printed in the United States of America

UNUP-1102
ISBN 92-808-1102-9

Library of Congress Cataloging-in-Publication Data

Key issues for mountain areas / edited by Martin F. Price, Libor F. Jansky, and
Andrei A. Iastenia.
    p. cm.
  Includes bibliographical references and index.
  ISBN 92-808-1102-9 (pbk.)
  1. Sustainable development. 2. Mountains. I. Price, Martin F. II. Jansky,
Libor. III. Iastenia, Andrei A. IV. Title.
HC79.E5K434 2004
338.9'27'09143—dc22   2004015893

Contents

List of tables and figures

Note on measurements

In this volume:

1 billion = one thousand million
$1   = 1 US dollar

Preface

The degradation of mountain ecosystems – home to 600 million people and the source of water for more than half the world's population – threatens to seriously worsen already existing global environmental problems, including floods, landslides, and famine. Climate change, pollution, armed conflict, population growth, deforestation, and exploitative agricultural, mining, and tourism practices, are among a growing list of problems confronting the "water towers of the world," prompting warnings that catastrophic flooding, landslides, avalanches, fires, and famines will become more frequent and that many unique animals and plants will disappear. Although several of the world's mountain areas are in relatively good ecological shape, many face accelerating environmental and cultural decline brought on, in part, by government and multilateral agency policies too often founded on inadequate research.

We know that the environment has steadily worsened in the last 30 years, despite the many actions society has taken since the first serious discussions took place in the 1970s. In addition, the impact of globalization is increasing the strain on the use of our natural resources. Sustainable development has slipped down the political totem pole and has become overshadowed by concerns for security and economic globalization. As a result, such issues as "trade and environment" or "how to manage globalization" have become seemingly insurmountable obstacles to achieving sustainable development.

The International Year of Mountains was an opportunity and invitation to the scientific community to foster better, more effective, support and development policies by improving the world's understanding of environmental and other problems facing mountain regions. "Cooperation" and "partnership" are the tools for developing, promoting, and implementing programmes, policies, and approaches towards (realistic) sustainable mountain development. In Johannesburg, the World Summit on Sustainable Development (WSSD) advanced a sense of how to actually implement sustainable development – not an easy task, but one essential to the ultimate well-being of both high- and lowlanders. The International Partnership for Sustainable Development in Mountain Regions, launched at the WSSD, will have a crucial role in this.

The following immediate and long-term policy suggestions have resulted from the UN agencies' long involvement in mountain issues and its networking with the world's mountain scholars:

- Strengthening of knowledge about the ecology and sustainable development of mountain ecosystems – more research and monitoring to identify knowledge gaps, needs, and constraints;
- Capacity development for mountain populations and minorities to counteract marginalization;
- Maintenance and development of cultural diversity;
- Holistic and interdisciplinary management schemes for environmental conservation and sustainable development;
- Dissemination of more realistic and accurate information through the mass media;
- Greater attention to urban aspects of mountain issues;
- Empowerment of local communities, especially women;
- More attention to conflicts and resulting destruction of mountain ecosystems and livelihoods;
- Promoting integrated watershed development and opportunities for alternative livelihoods.

The focus of *future research* should include the following: (1) coexistence between peoples having different cultures, languages, and social systems; (2) maintenance of peace and security; (3) human rights; (4) economic and social changes; (5) the proper use of both human and natural resources; (6) application of the results of science and technology in the interest of development of mountain regions; and (7) human values related to improvement of quality of life. It is important to provide assistance to enable participation in research, in order to increase the capacity of individuals and groups to contribute to the extension and application of relevant knowledge within interdisciplinary approaches.

The objective of any *capacity development* effort is to develop human potential to directly address any deterioration of human security and development conditions. It can be understood as an integral part of the research and policy studies, as an educational activity with interdisciplinary approach, and as an integral part of policy development and decision-making processes. The goals and objectives of capacity development are as follows: (1) to build a knowledge base and awareness that facilitates better decision-making; (2) to improve individual health, literacy, and other skills required to adapt to differing and changing circumstances; (3) to integrate laws, policies, and strategies that encourage sustainable development, including environmental integrity; (4) to improve management practices and techniques; (5) to foster institutions that promote and support partnerships and cooperative arrangements; (6) to develop appropriate infrastructure and technology to support sustainable development; and (7) to identify and promote sustainable financing mechanisms. This list of goals and objectives is not intended to be exhaustive.

Nevertheless, the term "capacity development" has become overused in recent years to the point where it has lost most of its meaning. Capacity development is the process whereby a community equips itself to undertake the necessary functions of governance and service provision in a sustainable fashion. This process must have the aim both of increasing access to resources and of changing the power relationships between the parties involved. The "community" may comprise a local government, a village-level committee, or even a central government department.

Each region features a complex array of strengths and problems, making it impossible to propose a generalized approach to mountain-related issues. It is possible to generalize, however, about the lack of information needed for effective policy formulation. The data on which policy makers rely often relate to mountain ranges in the developed world and are inappropriately applied to developing countries. Notions based on scant scientific data are often accepted as truths: for example, although serious problems exist in the Himalaya, massive deforestation has not occurred across the entire mountain system. Such misinformed assumptions have led to simplistic and, often, counter-productive remedies.

In addition to the necessity of gathering and sharing more and better data and information worldwide, there is an urgent need to strengthen capacity in the mountain regions of developing countries – capacity, for instance, in meteorology, hydrology, ecology, and soil sciences. Much firmer links to the human sciences – anthropology, social science, and human geography – should also be established. The management of mountain regions and watersheds in a way that embraces and integrates many sciences will be key to success. The promotion of opportunities for alternative livelihoods for mountain people in developing countries is essential; this should help to alleviate the poverty at the root of so many of their health and environmental problems.

In all these activities, however, it is necessary to recognize that the mountain minority people worldwide, who are among the poorest of the poor, are extremely rich in environmental understanding: their opinions and experiences need to be combined with scientific knowledge before a better understanding of mountain processes can be obtained. Cultural diversity, which is a prevailing feature of mountain life, must be considered as complementary to biological diversity if sustainable mountain development is to be achieved. The widespread conflicts in mountain regions – including conventional warfare, terrorism, guerrilla insurgency, and repression of minority peoples – must be tackled far more vigorously than hitherto. The management and utilization of the natural resources of mountains, especially water, must be undertaken in such a way that mountain people share in the benefits. Achievement of the equality of access to resources for both men and women also requires much greater attention.

The monitoring and collection of data is measurement, not research, as it leads to the accumulation of (scientific) information rather than to knowledge. True knowledge is based on understanding and leads not merely to being aware of a situation but also to knowing how to change or influence that situation – or, if that is impossible, how to adapt to the situation and to live with it.

Lastly, I would like to express my deep gratitude to all the organizations and people that have contributed to this publication, in particular UNEP. Without their continuous and strong support this publication would not have been possible.

Hans J.A. van Ginkel
Rector, United Nations University
UN Under-Secretary General

1

Introduction: Sustainable mountain development from Rio to Bishkek and beyond

Martin F. Price

Introduction

Mountains occupy 24 per cent of the global land surface (Kapos et al. 2000) and host 12 per cent of the global population (Huddleston et al. 2003). A further 14 per cent of the global population lives adjacent to mountain areas (Meybeck, Green, and Vörösmarty 2001); mountain people include not only remote, poor, and disadvantaged people and communities but also wealthy tourist communities and also urban centres within and close to the mountains – including megacities such as Mexico City and Jakarta. As sources of water, energy, and agricultural and forest products, and as centres of biological and cultural diversity, religion, recreation, and tourism, mountains are important for at least half of humanity (Messerli and Ives 1997).

These statistics show the global importance of mountains. Yet, just over a decade ago, the world's mountains were a topic of interest to a relatively small number of scientists, development experts, and decision makers, as well as mountaineers. The United Nations Conference on Environment and Development (UNCED), held in Rio de Janeiro in 1992, presented a unique opportunity to move mountains onto the global stage, through the inclusion of a specific chapter in Agenda 21, the plan for action endorsed at UNCED by the Heads of State or Government of most of the world's nations (Price 1998; Stone 2002). Chapter 13 of

Agenda 21 is entitled "Managing fragile ecosystems: sustainable mountain development," and includes two "programme areas":

- generating and strengthening knowledge about the ecology and sustainable development of mountain ecosystems;
- promoting integrated watershed development and alternative livelihood opportunities.

That chapter meant that, for the first time, mountains were accorded comparable priority in the global debate about environment and development with issues such as global climate change, desertification, and deforestation. In 1998, the UN General Assembly re-emphasized the importance of the world's mountains by declaring the year 2002 the International Year of Mountains (IYM).

At the global level, formal implementation of Chapter 13 began in 1993, when the UN Inter-Agency Committee on Sustainable Development appointed the Food and Agriculture Organization of the United Nations (FAO) as Task Manager for Chapter 13. In this role, FAO has convened an ad hoc Inter-Agency Group on Mountains (IAGM) which, in spite of its name, involves more than UN agencies. From the beginning, FAO recognized that diverse actors are involved in processes relating to the sustainable development of mountain areas. Consequently, FAO invited a number of non-governmental organizations (NGOs) to join the group, and they have participated in all seven meetings to date. Among the recommendations made by the IAGM at its first meeting was that national governments and NGOs should become directly involved in the implementation of Chapter 13. A key means to this end was a series of regional intergovernmental consultations, bringing together governments within the African, Asia-Pacific, European, and Latin America/ Caribbean regions in 1994–1996. In total, representatives of 62 countries and the European Union attended these meetings (Price 1999).

Parallel to this intergovernmental process, a non-governmental process took place. Its importance was underlined by the IAGM, recognizing that the process that had led to Chapter 13 – in contrast to many other chapters of Agenda 21 – had been driven by a relatively small number of academics and development experts, mainly from industrialized countries. In 1995, a global NGO consultation in Lima, Peru, brought together 110 participants from 40 countries. This meeting led to the establishment of the Mountain Forum – "a global network for mountain communities, environments, and sustainable development." The Mountain Forum has subsequently been organized through both global and regional structures and, at the end of 2003, comprised over 4,000 individual and 350 organizational members in more than 100 countries. Key means of information sharing include 15 discussion lists, electronic conferences, and an interactive website (www.mtnforum.org) with membership services, calendar of events, on-line library, and links to other networks (Taylor 2000).

In the five years following Rio, a number of countries established national-level or sub-national institutions concerned with the sustainable development of their mountain areas. Others, particularly in Europe, developed laws and policies effectively to this end (Price 1999; Villeneuve, Castelein, and Mekouar 2002; Villeneuve, Hofer, and McGuire, ch. 9, this volume). Many other related activities took place in various nations around the world, organized both at national and sub-national level and also by international organizations, particularly the FAO, the United Nations Educational, Scientific and Cultural Organization (UNESCO), and the United Nations University (UNU), all of which had long-standing activities in mountain areas. In 1995, the Global Environment Facility (GEF) identified mountain ecosystems as the subject of one of its ten operational programmes; by 2002, it had committed over US$620 million and leveraged about $1.4 billion of additional funding for at least 107 mountain-related projects in 64 countries (Walsh 2002).

It was in this context of a gathering international momentum of support for mountain areas that the participants in the international conference "Mountain Research – Challenges for the 21st Century," held in Bishkek, Kyrgyzstan in 1996, proposed that sustainable mountain development should be the theme of a forthcoming international year. This idea was proposed to the UN Economic and Social Council (ECOSOC) by the Kyrgyz Ambassador to the United Nations in 1997, resulting in a resolution, co-sponsored by 44 member countries, requesting the Secretary-General to undertake an exploratory process. At its subsequent session, ECOSOC adopted a resolution, co-sponsored by 105 member countries, recommending to the General Assembly that 2002 should be declared the International Year of Mountains (IYM). The outcome was that the UN General Assembly proclaimed, at its fifty-third session in 1998, in a resolution sponsored by 130 countries, that 2002 would be the IYM.

Sustainable (mountain) development: Definition and indicators

The term "sustainable mountain development" (SMD) appeared first in the title of Chapter 13 of Agenda 21. It includes two elements – (a) the concept of sustainable development and (b) mountains. The concept of sustainable development was introduced in the *World Conservation Strategy* (IUCN 1980). It became fashionable in the 1980s, particularly through the report of the World Commission on Environment and Development (WCED), or Brundtland Report, *Our Common Future*, which defined it as "development that meets the needs of the present without compromising the ability of future generations to meet their own needs" (WCED 1987). This is probably the most cited of a very large number of definitions: over a decade ago, Pezzey (1989) had identified 190, and the number has continued to increase (Murcott 1997). Another commonly used definition, agreed on by three of the major international organizations working in the field, is "development which improves the quality of life, within the carrying capacity of the earth's life support system" (IUCN/UNEP/WWF 1991).

Sustainable development was a keyword of UNCED and led to the establishment of the UN Commission on Sustainable Development. Yet debates about its meaning(s) continue, resulting inevitably from its appropriation by a wide range of authors and organizations in diverse cultures. However, most would agree that sustainable development is a process that aims at ensuring that current needs are satisfied while maintaining long-term perspectives regarding the use and availability of natural (and often other) resources into the long-term future, and considering the well-being of future generations.

Citing the title of Chapter 13, many meetings since UNCED, the documents deriving from them, and many projects started in the 1990s identified SMD as an objective. However, no attempt was made to define it until the end of the decade. If it is to be more than a vague goal, agreement on its meaning, and then on priorities and means for its implementation, is essential. In 1997, Sène and McGuire (1997) noted that "the concept of sustainable mountain development has taken on new meaning" since UNCED and stated that "[a] multi-sectoral, more comprehensive approach to addressing mountain development issues is a relatively new concept, but one whose time has come." They contrasted this multi-sectoral approach with past approaches to the problems and needs of mountain areas, which had largely been implemented within a sectoral context. They also noted the large number of themes addressed at the various regional intergovernmental consultations on SMD (Backmeroff, Chemini, and La Spada 1997; Banskota and Karki 1995; ILRI 1997; Mujica and Rueda 1996) and summarized by Price (1999). Although all of these documents provide long lists of issues that are intended to contribute (or in some way are related) to SMD, they are not prioritized – which is appropriate, given the very different characteristics of the world's diverse mountain regions, even on one continent.

Another key issue is the scale at which SMD should be implemented. For instance, one village may be able to develop a strategy for its own future that appears to be viable in the long term, yet this may have side-effects that are unsustainable for neighbouring or downstream communities. Along the many mountain ranges that form boundaries between countries and regions, there are particular needs for transboundary cooperation in SMD, given that ecological and societal processes and structures span these boundaries. The development of cooperative regional approaches is also important within the mountain massifs that are now divided between two or more nation-states but have long-established cultural and economic identities, distinct from adjacent lowlands in these states (Burhenne, ch. 10, this volume). In conclusion, it is probably best not to propose a precise definition of sustainable mountain development but to recognize that it is "a regionally-specific process of sustainable development that concerns both mountain regions and populations living downstream or otherwise dependent on these regions in various ways" (Price and Kim 1999).

The objectives of this process vary according to the size of the area concerned and are likely to shift over time. However, to assist in project development and wider planning and to evaluate success, indicators are needed. Various indicators have been proposed. At a global level, as part of an exercise using the pressure-state-response framework (OECD 1993), FAO (1996) proposed that the key pressure indicator is the population of mountain areas, to be measured in terms of population density, growth, and migration. Proposed key state indicators were, first, the welfare of mountain populations (to be measured in terms of nutritional anthropometry) and, second, qualitative assessment of the condition and sustainable use of natural resources in mountain areas. The latter indicator is a composite of four sub-indices used to describe the state of the natural-resource base of a watershed – namely, extent of protection of soil, area of "hazard" zones, extent of degraded land, and an indication of productivity. Other proposals have been made by Rieder and Wyder (1997), who (like many authors) suggest that sustainability should be measured in terms of three sets of indicators – ecological, economic, and social. Recognizing that indicators need to be tailored to specific circumstances, they discuss issues relating to economic, ecological, and social indicators for five mountain study areas – namely Bhutan, Encañada (Peru), Pays d'Enhaut (Switzerland), North Ossetia (Russia), and Puka (Albania). Finally, five European countries (Bulgaria, Hungary, Romania, Slovenia, Switzerland) suggested indicators of SMD in documents submitted to the second session of the European intergovernmental consultation in 1996.

Even at a regional or continental scale, agreement on priorities for SMD and how they should be measured will not be simple, as shown by a survey of key respondents working in governmental, non-governmental, and scientific organizations in 30 European countries (Price and Kim

1999). Using a set of 36 possible indicators derived from meetings on SMD in Europe, those authors found that, for all respondents, ecological priorities ranked higher than socio-political or economic priorities. However, there were two highly ranked socio-political variables: these were the empowerment of mountain communities and the need for education and training in conservation and development. Comparing respondents from "western" Europe with those from central/eastern Europe, they found that those in the latter region placed greater emphasis on ecological indicators. The greatest similarities were with regard to socio-political variables, implying a common interest in the more equitable provision of benefits to people in mountain areas, in order to reduce marginality and ensure the long-term survival of populations in these areas. Finally, comparing employees of government with those of NGOs and independent scientific organizations, the most significant differences were found: generally, the latter group ranked ecological issues more highly than socio-political or economic issues. Two of the most significant differences were with regard to (a) compensation for sustainable management of mountain ecosystems by downstream populations and (b) the creation of new livelihood opportunities. Interestingly, these two issues were seen as more important by the government employees, perhaps implying that they are more radical than suggested by the priorities of the organizations for which they work. Similarly, workshops of "specialists" and local stakeholders in the Cairngorms of Scotland found greater agreement between the two groups with regard to indicators of "natural capital" than for those relating to economic and social and political factors (Bayfield, McGowan, and Fillat 2000). Although there has been no comparable research in other parts of the world, it appears desirable that indicators for SMD should be appropriate to the region of concern and based on data that are measurable, available, easily understood, and meaningful (Rieder and Wyder 1997). However, as shown by Parvez and Rasmussen (ch. 5, this volume), such data are often not available at a fine enough scale.

The International Year of Mountains: Objectives and activities

The mission statement of the IYM, developed by FAO in its role as Lead Agency for the Year, was to "promote the conservation and sustainable development of mountain regions, thereby ensuring the well-being of mountain and lowland communities." As stated in the concept paper for the IYM, it "should provide an opportunity to initiate processes that eventually advance the development of mountain communities, and act as a 'springboard' or catalyst for long-term, sustained, and concrete action" (FAO 2000).

The IYM represented a unique opportunity to raise awareness, across society as a whole, of the manifold values of mountain regions and the fragility of their resources, building on the IYM motto "We are all mountain people." Around the world, diverse media – postage stamps, newspapers, magazines, radio, television, the Internet – featured mountain issues. Many reports and books on mountain issues were published (e.g. Blyth et al. 2002; Körner and Spehn 2002; Royal Swedish Academy of Sciences 2002). All these means raised the awareness of innumerable people with regard to the diverse values of mountains at all scales – an investment in their future, as the IYM must not be regarded as a "one-off" but as a unique year in the process of fostering SMD.

National committees

During the planning of the IYM, it was recognized that one measure of success would be the extent to which it contributed to establishing effective programmes, projects, and policies. Although this requires participation at all levels, from individual villages and NGOs to international organizations, the greatest efforts need to come from those working at the national level to achieve SMD. Thus, as for other International Years, great emphasis was given to the establishment of national committees for the IYM. By the end of 2002, with the support of FAO, 78 countries had established such national committees or similar mechanisms. Although most of these were led by a government agency, many included representatives of mountain people, grass-roots organizations, NGOs, the private sector, research institutions, UN agencies, national government agencies, and decentralized authorities. In some countries, the national IYM committee was the first national mechanism for the sustainable development of mountains and the first opportunity to implement a holistic approach to mountains.

During the IYM, new mountain laws were passed in Kyrgyzstan and drafted in Morocco and Romania; in Korea, the Korea Forest Service (which took the lead for the IYM) prepared a Forest Management Law that was passed at the end of December. National mountain strategies and plans were developed in Madagascar, Spain, and Turkey (Villeneuve, Castelein, and Mekouar 2002; Villeneuve, Hofer, and McGuire, ch. 9, this volume). A number of national committees may disappear; nevertheless, all provided opportunities for dialogue. All have been encouraged to continue to operate – and it is anticipated that many will do so in order to help develop and implement sustainable development strategies, policies, and laws designed to respond to the specific needs, priorities, and conditions of the mountain areas of their respective countries.

Meetings

As with any International Year, the IYM was marked by numerous meetings and other events, on almost every possible theme relating to mountains – mountain women, children, water, mining, war, forests, biodiversity, arts.... All were important because they brought together many people who would otherwise never have met, leading to increased understanding both of issues and of others' viewpoints, and raising awareness in various ways. Key regional meetings included the Seventh Alpine Conference, at which the vital decision on the location of the Secretariat of the Alpine Convention was made (see Burhenne, ch. 10, this volume); two meetings that accelerated the process towards a Carpathian Convention (Angelini, Egerer, and Tommasini 2002), leading to its signature in May 2003; and the ninth session of the African Ministerial Conference on the Environment in Uganda in July 2002, which produced the Kampala Declaration on the Environment for Development.

Eight major global meetings were associated specifically with the IYM (table 1.1). Four of these (in India, Bhutan, Peru, and Ecuador) specifically addressed the needs and interests of mountain people – respectively, children, women, indigenous people, and mountain populations. Two (both in Switzerland) addressed various aspects of development, particularly with regard to communities and agriculture, the latter linking Chapter 13 of Agenda 21 with Chapter 14 on sustainable agriculture and rural development. The "High Summit" was a truly global event, with simultaneous events on four continents bringing together mountain people, scientists, and representatives of NGOs, UN agencies, and the media through internet and videoconference technology.

All of these meetings produced final documents (see www.mtnforum. org) which fed into the final global event of the IYM, the Bishkek Global Mountain Summit held in Kyrgyzstan, which produced the Bishkek Mountain Platform (BMP) (Appendix A). This formulates recommendations for concrete action towards sustainable mountain development, providing guidance to governments and others on how to improve the livelihoods of mountain people, protect mountain ecosystems, and use mountain resources more wisely. The BMP was circulated at the fifty-seventh session of the UN General Assembly later in 2002, leading to the adoption of a resolution which, inter alia, designated 11 December as International Mountain Day and encouraged the international community to organize, on this day, events at all levels to highlight the importance of sustainable mountain development (Appendix B).

The Mountain Partnership

One key outcome of the IYM was the International Partnership for Sustainable Development in Mountain Regions, or "Mountain Partnership." Its outline was developed by the Swiss Government, FAO, and the United Nations Environment Programme (UNEP) during the fourth Preparatory Meeting for the World Summit on Sustainable Development (WSSD) in Bali. The Partnership was launched at the WSSD in Johannesburg; as at UNCED, ten years before, the meeting's final document specifically refers to mountains – this time, in paragraph 42 of the Plan of Implementation. The primary purpose of the Partnership is to address the second of the two goals of Chapter 13 of Agenda 21 – to improve livelihoods, conservation, and stewardship across the world's mountains. It is conceived as a mechanism for improving, strengthening, and promoting greater cooperation between all mountain stakeholders. It will be aimed at clearly agreed goals, its operations will be based on commitments made by partners, and its implementation will be supported through better linkages between institutions and improved monitoring systems.

The Partnership was one of the main topics of discussion at the Bishkek Global Mountain Summit. In the BMP, the participants welcomed the offer of the FAO to host its secretariat and to bring the IAGM to its service. They also called on UNEP to ensure environmentally sound management in mountain regions – in particular, in developing countries – by strengthening environmental networking and assessments, facilitating regional agreements, and encouraging public–private-sector cooperation. In addition, other UN agencies, multilateral development banks, and other international organizations and states were recognized as key players. Both the actual structure and function of the Partnership were developed during 2003, through a process including an electronic consultation organized by Mountain Forum, discussion at the annual meeting of the UN Commission on Sustainable Development, and a meeting in Merano, Italy. By December 2003, 40 countries, 15 intergovernmental organizations, and 38 other organizations ("major groups") had expressed their interest in actively taking part.

Introduction to this volume

During the preparation of the Bishkek Global Mountain Summit (BGMS), the International Advisory Board for the BGMS recognized the need for a series of background papers around which the meeting would be structured. Following the identification of the themes from

Table 1.1 Global meetings associated with the International Year of Mountains (IYM)

Title	Dates; location	Participants	Organizers	Outcome
World Mountain Symposium 2001: Community Development between Subsidy, Subsidiarity and Sustainability	30 September–4 October 2001; Interlaken, Switzerland	150 participants from 56 countries	Swiss Agency for Development and Cooperation, Centre for Development and Environment, University of Berne	Proceedings, CD
High Summit 2002: International Conference around the Continents' Highest Mountains	6–10 May 2002; Mendoza, Argentina; Nairobi, Kenya; Kathmandu, Nepal; Milan and Trento, Italy	Mountain people, scientists, representatives of NGOs, UN agencies, and the media	Italian National Committee for the IYM	Recommendations for action on five cornerstones of mountain development: water, culture, economy, risk, and policy
International Conference of Mountain Children	15–23 May 2002; Uttaranchal, India	Children from 13 to 18 years of age from over 50 countries	Research Advocacy and Communication in Himalayan Areas	Recommendations for the Bishkek Mountain Platform, Internet-based Mountain Children's Forum
2nd International Meeting of Mountain Ecosystems, "Peru, country of mountains towards 2020: water, life and production"	12–14 June 2002; Huaraz, Peru	300 participants from 16 countries, especially indigenous people from Peru, Ecuador, and the Himalayas	National Committee of Peru for the IYM	Huaraz Declaration

Event	Date; Location	Participants	Organizer	Output
International Conference on Sustainable Agriculture and Rural Development in Mountain Regions	16–20 June 2002; Adelboden, Switzerland	200 people from 50 countries	Swiss Federal Office for Agriculture	Adelboden Declaration
Second World Meeting of Mountain Populations	17–22 September 2002; Quito, Ecuador	Representatives of 115 countries	World Mountain Peoples Association, El Centro de Investigación de los Movimientos Sociales del Ecuador	Quito Declaration: Draft Charter for World Mountain People
Celebrating Mountain Women	1–4 October 2002; Thimphu, Bhutan	250 participants from 35 countries: civil society, NGOs, media, academia, development agencies, donors	International Centre for Integrated Mountain Development and Mountain Forum	Thimphu Declaration
Bishkek Global Mountain Summit	28 October–1 November 2002; Bishkek, Kyrgyzstan	Over 600 people from 60 countries	Government of Kyrgyzstan, with assistance from UNEP	Bishkek Mountain Platform

among the great variety relevant to SMD, and recognizing the existence of key syntheses, such as those by Messerli and Ives (1997) and Funnell and Parish (2001), the first drafts of the papers were prepared by international experts and then considered in an electronic consultation (e-consultation) organized by the Mountain Forum. During this process, over a period of two weeks, each paper was posted on the Mountain Forum website. Participants in the e-consultation were invited to comment by email on the papers – with some comments leading to further discussion – and to contribute case studies for possible incorporation in the papers. Following the e-consultation, the papers were revised and submitted to peer review by other international experts. The final versions were presented at the BGMS. Subsequently, they were again revised and updated to form the chapters of the present volume.

Chapter 2, by Iyngararasan and colleagues, addresses the diverse challenges of mountain environments and their relevance for the global population. Attention is given to issues including the key values of mountains as "water towers" (Bandyopadhyay et al. 1997; Liniger and Weingartner 1998), the high frequency of natural hazards (Hewitt 1997), the potential impacts of climate change (Price and Barry 1997), and regional issues such as regional haze and desertification. A number of existing initiatives and best practices are described, and future needs discussed. Chapter 3, by Kohler and colleagues, addresses access, communications, and energy (Schweizer and Preiser 1997) – three sets of key issues for the development of mountain regions and their integration in wider economies. They recognize that the development of access, communications, and energy has often been driven by the needs of lowland populations; they propose that, in future, mountain people should be directly involved in such development, bringing shared benefits and using appropriate technologies, which often build on the long-term experience and institutions of mountain people.

The links between mountain and lowland regions are explicitly considered in chapter 4 by Koch-Weser and Kahlenborn, in the context of economic and policy instruments. They critically review a number of environmental services agreements, designed to ensure that mountain people are fairly compensated for services they provide to downstream communities and enterprises. The development of such market-based mechanisms is a key element of SMD; this chapter addresses such mechanisms specifically in the mountain context, building on other work such as that focussing on the environmental services provided by forest ecosystems (Pagiola, Bishop, and Landell-Mills 2002). The criteria for developing effective mechanisms and agreements will be of use in many mountain regions.

Chapter 5, by Parvez and Rasmussen, addresses questions of disparities between mountain and non-mountain countries, and between mountain and lowland regions. The chapter shows that, despite the extensive literature describing poverty in mountain regions (Ives 1997), national and sub-national statistics – the latter principally from South Asia and China – do not show that mountain regions necessarily have a weaker development performance: performance appears to be more closely related to national trends, and strong national economies are important in supporting the development of mountain regions through policy and financial means. They conclude that a "sustainable livelihoods" approach may be more appropriate for understanding mountain development issues and suggesting appropriate policies. In this context, the issues addressed by Brewer Lama and Sattar in chapter 6 are highly relevant. Mountain regions are centres of biological and cultural diversity (Bernbaum 1997; Grötzbach and Stadel 1997; Jeník 1997) and these are fundamental bases for tourism, which has become the economic mainstay of many mountain communities (Price, Moss, and Williams 1997); however, tourism can be only one element of SMD. A number of principles and necessary actions for sustainable mountain tourism are presented, complemented by brief descriptions of best practices from around the world.

Chapter 7, by Pratt, continues the discussion on sustainability, recognizing two general approaches – local, drawing from traditional cultures, and linked, in which mountain and downstream populations are linked in various ways, as described by previous authors. A number of types of institutional arrangements are described; their appropriateness in any particular region depends on the interactions of two sets of criteria – local/linked economies and the values of natural resources and environmental services. In all cases, democratic and decentralized institutions are important, but their development and application depend on the existence of appropriate incentives. In this vein, Starr addresses issues relating to conflict and peace in mountain societies in chapter 8. A significant proportion of conflicts around the world occur in mountain areas (Libiszewski and Bächler 1997). These conflicts typically derive from problems of social and economic breakdown whose roots are largely outside the mountain areas themselves. Returning to themes addressed by many of the previous authors, particularly Parvez and Rasmussen, the conclusion is that the resolution of conflict requires attention to people, especially their security and economic development.

Laws and policies are essential elements of SMD, although it must be recognized that their existence is only the prelude to their effective implementation. Chapter 9, by Villeneuve and colleagues, describes the diversity of laws, policies, and institutions that explicitly address mountain issues in countries around the world. As mentioned above, their number has increased during and since the IYM. However, many issues relating to mountain regions are transnational: ecosystems straddle national boundaries; water, air, fires, animals, diseases, and people – among others – cross them. Consequently, international agreements for mountain regions are important. In chapter 10, Burhenne provides the principles for such agreements, and briefly describes their application, especially with regard to the Alpine Convention.

The concluding chapter, by Messerli and Bernbaum, addresses the roles of culture, education, and science for SMD. All have key roles to play. Most mountain cultures have long traditions, deeply rooted in the places where they have developed; however, there are significant needs to find ways to draw on long-standing strengths in adapting to a rapidly changing world. Traditional knowledge can be of considerable benefit in this context and should be explicitly considered in the development and implementation of education, at all levels, which provides the tools necessary for mountain people to move towards SMD during the twenty-first century and beyond. Modern technologies may be of particular benefit: as Kohler and colleagues point out in chapter 3, many mountain people have better access to the wider world through information and communications technologies (ICT), such as mobile telephones and internet connections, than through traditional means, such as roads and railways. In this and many other ways, the diverse branches of science have vital roles to play in SMD. Informed science is essential for policy-making and, in an increasingly complex world, interdisciplinary and transdisciplinary approaches are essential.

Considered together, and particularly in conjunction with the chapters in Messerli and Ives (1997), the chapters in this book underline the fact that the world's mountain regions are inextricably woven into a global fabric of interlinked markets, institutions, and policies within a biosphere that is experiencing rapid change. In other words, mountain environments (and the billions of people who depend on them) are affected by all the ecological and societal processes of global change. This has been recognized through the development of the Mountain Research Initiative (MRI) (Becker and Bugmann 2001) which, within the major global research programmes on global change, attempts to develop a coherent understanding of all these processes in order to contribute to SMD both regionally and globally. The MRI is one example of a partnership and will contribute to the Mountain Partnership. The strengthening of existing partnerships (and the development of new ones) is particularly appropriate in mountain regions, as cooperation is one of the distinguishing characteristics of mountain societies: in such uncertain environments, it has long been recognized that sharing resources and working together is essential for long-term survival. The integration of mountain areas into regional and global economies has often reduced the effectiveness of such cooperative structures as external interests come to dominate. The chapters in this volume show not only many of the challenges but also that partnerships, both within mountain regions and between stakeholders in mountain regions and those outside, are essential for sustainable mountain development.

REFERENCES

Angelini, P., H. Egerer, and D. Tommasini (eds). 2002. *Sharing the experience: Mountain sustainable development in the Carpathians and the Alps.* Bolzano: Accademia Europea Bolzano.

Backmeroff, C., C. Chemini, and P. La Spada (eds). 1997. European Intergovernmental Consultation on Sustainable Mountain Development. Proceedings of the Final Trento Session. Trento: Provincia Autonoma di Trento.

Bandyopadhyay, J., J.C. Rodda, R. Kattelmann, Z.W. Kundzewicz, and D. Kralmer. 1997. "Highland waters – a resource of global significance." In: B. Messerli and J.D. Ives (eds) *Mountains of the world: A global priority.* Carnforth: Parthenon.

Banskota, M., and A.S. Karki (eds). 1995. *Sustainable development of fragile mountain areas of Asia.* Kathmandu: International Centre for Integrated Mountain Development.

Bayfield, N.G., G.M. McGowan, and F. Fillat. 2000. "Using specialists or stakeholders to select indicators of environmental change for mountain areas in Scotland and Spain." *Oecologia Montana* Vol. 9.

Becker, A., and H. Bugmann. 2001. *Global change and mountain regions: The Mountain Research Initiative.* Stockholm: IGBP Secretariat.

Bernbaum, E. 1997. "The spiritual and cultural significance of mountains." In: B. Messerli and J.D. Ives (eds) *Mountains of the world: A global priority.* Carnforth: Parthenon.

Blyth, S., B. Groombridge, I. Lysenko, L. Miles, and A. Newton. 2002. *Mountain Watch: Environmental change and sustainable development in mountains.* Cambridge: UNEP World Conservation Monitoring Centre.

Food and Agriculture Organization of the United Nations (FAO). 1996. "Criteria and indicators for sustainable mountain development." Internal Report. Rome: FAO.

Food and Agriculture Organization of the United Nations (FAO). 2000. *International Year of Mountains: Concept paper.* Rome: FAO.

Funnell, D., and R. Parish. 2001. *Mountain environments and communities.* London: Routledge.

Grötzbach, E., and C. Stadel. 1997. "Mountain peoples and cultures." In: B. Messerli and J.D. Ives (eds) *Mountains of the world: A global priority.* Carnforth: Parthenon.

Hewitt, K. 1997. "Risk and disaster in mountain lands." In: B. Messerli and J.D. Ives (eds) *Mountains of the world: A global priority.* Carnforth: Parthenon.

Huddleston, B., E. Ataman, P. de Salvo, M. Zanetti, M. Bloise, J. Bel, G. Francheschini, and L. Fè d'Ostiani. 2003. *Towards a GIS-based analysis of mountain environments and populations.* Working Paper no. 10. Rome: FAO.

International Livestock Research Institute (ILRI). 1997. "Sustainable Development in Mountain Ecosystems of Africa." Proceedings of the African Intergovernmental Consultation on Sustainable Mountain Development. Addis Ababa: ILRI.

International Union for the Conservation of Nature (IUCN). 1980. *World conservation strategy.* Gland: IUCN.

International Union for the Conservation of Nature/United Nations Environment Programme/World Wildlife Fund (IUCN/UNEP/WWF). 1991. *Caring for the Earth: A strategy for sustainable development.* Gland: IUCN.

Ives, J.D. 1997. "Comparative inequalities – mountain communities and mountain families." In: B. Messerli and J.D. Ives (eds) *Mountains of the world: A global priority.* Carnforth: Parthenon.

Jeník, J. 1997. "The diversity of mountain life." In: B. Messerli and J.D. Ives (eds) *Mountains of the world: A global priority.* Carnforth: Parthenon.

Kapos, V., J. Rhind, M. Edwards, M.F. Price, and C. Ravilious. 2000. "Developing a map of the world's mountain forests." In: M.F. Price and N. Butt (eds) *Forests in sustainable mountain development: A state-of-knowledge report for 2000.* Wallingford: CAB International.

Körner, C., and E.M. Spehn (eds). 2002. *Mountain biodiversity: A global assessment.* New York: Parthenon.

Libiszewski, S., and G. Bächler. 1997. "Conflicts in mountain areas – a predicament for sustainable development." In: B. Messerli and J.D. Ives (eds) *Mountains of the world: A global priority.* Carnforth: Parthenon.

Liniger, H., and R. Weingartner. 1998. "Mountains and freshwater supply." *Unasylva* Vol. 49, No. 195.

Messerli, B., and J.D. Ives (eds). 1997. *Mountains of the world: A global priority.* Carnforth: Parthenon.

Meybeck, M., P. Green, and C. Vörösmarty. 2001. "A new typology for mountains and other relief classes: An application to global continental water resources and population distribution." *Mountain Research and Development* Vol. 21.

Mujica, E., and J.L. Rueda (eds). 1996. *El Desarollo Sostenible de Montañas en América Latina.* Lima: CONDESAN/CIP.

Murcott, S. 1997. "Definitions of sustainable development." [www.sustainableliving.org/appen-a.htm].

Organization for Economic Cooperation and Development (OECD). 1993. *OECD core set of indicators for environmental performance reviews.* OECD Environment Monographs No. 83. Paris: OECD.

Pagiola, S., J. Bishop, and N. Landell-Mills (eds). 2002. *Selling forest environmental services: Market-based mechanisms for conservation and development.* London: Earthscan.

Pezzey, J. 1989. *Economic analysis of sustainable growth and sustainable development.* Environmental Working Paper No. 15. Washington, D.C.: World Bank.

Price, M.F. 1998. "Mountains: Globally important ecosystems." *Unasylva* Vol. 49, No. 195.

Price, M.F. 1999. *Chapter 13 in action 1992–97 – A Task Manager's report.* Rome: FAO.

Price, M.F., and R.G. Barry. 1997. "Climate change." In: B. Messerli and J.D. Ives (eds) *Mountains of the world: A global priority.* Carnforth: Parthenon.

Price, M.F., and E.G. Kim. 1999. "Priorities for sustainable mountain development in Europe." *International Journal of Sustainable Development and World Ecology* Vol. 6.

Price, M.F., L.A.G. Moss, and P.W. Williams. 1997. "Tourism and amenity migration." In: B. Messerli and J.D. Ives (eds) *Mountains of the world: A global priority.* Carnforth: Parthenon.

Rieder, P., and J. Wyder. 1997. "Economic and political framework for sustainability of mountain areas." In: B. Messerli and J.D. Ives (eds) *Mountains of the world: A global priority.* Carnforth: Parthenon.

Royal Swedish Academy of Sciences. 2002. "The Abisko Document: Research for mountain area development." *Ambio Special Report Number 11.* Stockholm: Royal Swedish Academy of Sciences.

Schweizer, P., and K. Preiser. 1997. "Energy resources for remote highland areas." In: B. Messerli and J.D. Ives (eds) *Mountains of the world: A global priority.* Carnforth: Parthenon.

Sène, E.H., and D. McGuire. 1997. "Sustainable mountain development: Chapter 13 in action." In: B. Messerli and J.D. Ives (eds) *Mountains of the world: A global priority.* Carnforth: Parthenon.

Stone, P.B. 2002. "The fight for mountain environments." *Alpine Journal* Vol. 107.

Taylor, D.A. 2000. "Mountains on the move." *Américas* Vol. 52, No. 4.

United Nations. 1992. *Agenda 21: Chapter 13: Managing fragile ecosystems: Sustainable mountain development.* UN Doc. A/CONF. 151/26 (Vol. II), 13 August. Washington D.C.: United Nations.

Villeneuve, A., A. Castelein, and M.A. Mekouar. 2002. *Mountains and the law: Emerging trends.* Rome: FAO.

Walsh, S. (ed.). 2002. *High priorities: GEF's contribution to preserving and sustaining mountain ecosystems.* Washington D.C.: Global Environmental Facility.

World Commission on Environment and Development (WCED). 1987. *Our common future.* Oxford: Oxford University Press.

2

The challenges of mountain environments: Water, natural resources, hazards, desertification, and the implications of climate change

Mylvakanam Iyngararasan, Li Tianchi, Surendra Shrestha, P.K. Mool, Masatoshi Yoshino, and Teiji Watanabe

Summary

Mountain ecosystems harbour a wide range of significant natural resources and play a critical role in ecological and economic processes worldwide. Deforestation, landslides, land degradation, desertification, and glacier lake outburst flooding (GLOF) are some of the common environmental issues in the mountain regions. The major challenge currently faced by the mountain environment is the escalation of these issues through atmospheric changes.

Mountain systems are particularly sensitive to climate change. Global average surface temperatures increased by $0.6 \pm 0.2°C$ during the twentieth century; the global average surface air temperature is projected to increase by 1.4–5.8°C by 2100 relative to 1990. Analysis of the temperature trend in the Himalaya and its vicinity shows that temperature increases are greater in the uplands than lowlands. Regional changes in climate have already affected diverse physical and biological systems in many parts of the mountain regions. Such trends may be exacerbated by other atmospheric changes, such as regional haze.

Mountains are the water towers of the world. Major trends in recent years include unpredicted river flows, frequent floods, droughts, and crop failures. The management and protection of water resources have reached a crucial period. The shrinkage of glaciers is an ongoing trend, linked to the serious hazard of GLOF. Thousands of lives are lost every year in mountains and adjoining regions, owing, in particular, to the high frequency of natural hazards, some of which are restricted to mountain areas and others being more frequent in these areas; all are major constraints to sustainable development. Climate change also increases the vulnerability of mountain environment to desertification, leading to a vicious cycle of poor vegetation and poor soil.

Mountain issues cannot be tackled by the mountain community or by individual countries alone, especially because of the emerging challenges from the atmospheric issues. Partnerships between existing institutions and programmes concerned with mountain and atmospheric issues are vital to tackle the issues. Existing international initiatives and regional agreements should be adopted, recognizing the need to work together. Capacity building to strengthen the scientific base of knowledge and the establishment of monitoring and early warning systems are essential to tackle the challenges.

The issues

Mountains and uplands cover about 24 per cent of the Earth's surface, and influence most of the planet. The most important influence is on the hydrological cycle. Mountains act as barriers to the flow of moisture-bearing winds and control precipitation in neighbouring regions. For example, the Himalaya is of fundamental importance to the occurrence of the monsoon in northern India and to the continental arid conditions in Central Asia.

Until mountain areas were integrated into industrial economies, upland–lowland interactions were based primarily on the needs of upland communities. The transactions involved the bare essentials. As mountain populations and accessibility to mountain areas have increased, mountain resources and people have moved downhill while environmental degradation and social ills have climbed uphill. Deforestation, landslides, land degradation, desertification, and GLOF are key environmental issues in mountain regions, which are particularly susceptible to natural hazards. Atmospheric changes are now a major challenge for those concerned with mountain environments: emerging issues are climate change and emissions of aerosols and acidifying substances. These processes result from emissions from the industrial, transport, and domestic sectors. Figure 2.1 shows emission estimates for carbon dioxide (CO_2) and sulphur dioxide (SO_2) for mountain regions.

In this chapter we attempt to analyse the processes of climate change

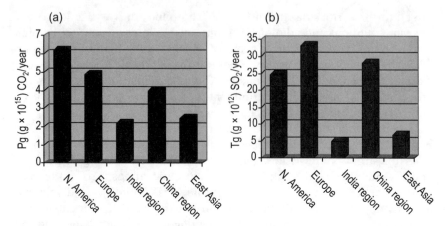

Figure 2.1 Estimated man-made (a) CO_2 and (b) SO_2 emissions. (India region: Bangladesh, Maldives, Sri Lanka, Myanmar, Nepal, Pakistan; China region: Cambodia, Viet Nam, Laos, Mongolia, North Korea; East Asia: Japan, South Korea, Indonesia, Malaysia, the Philippines, Thailand.)
Source: UNEP and C4 (2002); Van Aardenne et al. (2001)

and other atmospheric issues and their implications on mountain environments, with a particular focus on water, natural resources, hazards, and desertification.

Knowledge

Climate change

Since industrialization, human activities have resulted in steadily increasing concentrations of the greenhouse gases – particularly CO_2, methane (CH_4), chlorofluorocarbons (CFCs), and nitrous oxides (NO_x) – in the atmosphere. As these gases absorb some of the radiation emitted by the Earth rather than allowing it to pass through the atmosphere to space, there is general consensus that the Earth's atmosphere is warming. The third assessment report of the Intergovernmental Panel on Climate Change (IPCC 2001) concludes that global average surface temperatures have increased by $0.6 \pm 0.2°C$ over the twentieth century and that, for the range of scenarios developed, the global average surface air temperature is projected to increase by 1.4–5.8°C by 2100 relative to 1990.

Changes in temperature have not been consistent across the globe: mean air temperatures have increased more at higher than at lower latitudes. Equally, analysis of records from about 5,400 stations around the world for the period 1951–1989 has shown that monthly mean daily maximum temperature has been increasing at a rate of 0.88°C/100 years (Diaz and Bradley 1997). Similarly, an analysis of temperature trends in the Himalaya and its vicinity from 1977 to 1994 (Shrestha et al. 1999) shows that increases in temperature have been greater in the uplands than the lowlands. Du (2001) also found a clear trend of increasing temperature at most stations on the Tibetan Plateau in autumn and winter. Increases in daily minimum temperature have been mainly in winter, and have been greater than those for maximum temperature (which have been mainly in summer). Thus, a decrease in daily temperature range is clear, except in summer. The linear trend of the warming for annual mean air temperature is 0.26°C per decade above 4,000 m, but is relatively small (0.11°C per decade) below 3,000 m.

In Japan, there has been a clear increase in the mean air temperature in winter (December, January, and February) and summer (June, July, and August) at six Japanese mountain regions from 1971 to 2000 (Yoshino 2002). The greatest increases are in winter on Mounts Ibuki, Fuji, and Nikko, where the winter monsoon passes over the central part of Honshu. Thus, although the Japanese mountains show a tendency of temperature increase similar to those of Europe and Asia in general (Diaz and Bradley 1997), their rate of temperature increase is relatively greater because of the winter monsoon, which prevails strongly with colder air flows at the high troposphere over East Asia in winter.

Such regional changes in climate have already affected diverse physical and biological systems in many parts of the world. Shrinkage of glaciers, thawing of permafrost, late freezing and earlier break-up of ice on rivers and lakes, poleward and altitudinal shifts of plant and animal species, declines of some plant and animal populations, and earlier emergence of insects have been observed (IPCC 2001).

Climate influences weathering processes, erosion, sediment transport, and hydrological conditions. It also affects the type, quantity, quality, and stability of vegetation cover and, thereby, biodiversity. Mountain systems are particularly sensitive to climate changes: minor changes in climate can produce significant regional or larger-scale effects. In particular, marginal environments are under high stress: small changes in water availability, and floods, drought, landslides, and late frosts, can have drastic effects on agricultural economies.

Table 2.1 provides a summary of potential climate-change effects that are closely linked to mountain environments in different regions; more

Table 2.1 Climate change impacts by region

Region	Adaptive capacity, vulnerability, and key concerns
Africa	Major rivers are highly sensitive to climate variation; average runoff and water availability would decrease in Mediterranean and southern countries. Desertification would be exacerbated by reduction in average annual rainfall, runoff, and soil moisture, especially in southern, North, and West Africa. Increase in drought, floods, and other extreme events would add to stresses on water resources, food security, human health, and infrastructure, and would constrain development.
Asia	Extreme events (including floods, droughts, forest fires, and tropical cyclones) have increased in temperate and tropical Asia. Increased intensity of rainfall would increase flood risks in temperate and tropical Asia. Climate change would exacerbate threats to biodiversity due to land-use and land-cover change and population pressure in Asia. Poleward movement of the southern boundary of the permafrost zones of Asia would result in a change of thermokarst and thermal erosion with negative impacts on social infrastructure and industries.
Europe	Summer runoff, water availability, and soil moisture are likely to decrease in southern Europe, and would widen the difference between the north and drought-prone south. Half of the alpine glaciers and large permafrost areas could disappear by the end of the 21st century. River flood hazard will increase across much of Europe. Upward and northward shift of biotic zones will take place. The loss of important habitats would threaten some species.
Latin America	Loss and retreat of glaciers would adversely impact runoff and water supply in areas where glacier melt is an important water source. Floods and droughts would become more frequent, with floods increasing sediment loads and degrading water quality in some areas. The geographical distribution of vector-borne infectious disease would expand poleward and to higher elevations, and exposures to diseases such as malaria, dengue fever, and cholera would increase.
North America	Snowmelt-dominated watersheds in western North America would experience earlier spring peak flows, reduction in summer flows, and reduced lake levels. Unique natural ecosystems such as prairie wetlands, alpine tundra, and cold-water ecosystems would be at risk and effective adaptation is unlikely. Vector-borne diseases (including malaria, dengue fever, and Lyme disease) might expand their range.

Table 2.1 (cont.)

Region	Adaptive capacity, vulnerability, and key concerns
Polar	Climate change here is expected to be among the greatest and most rapid of any region on the Earth; it would have major physical, ecological, sociological, and economic impacts, especially in the Arctic, Antarctic Peninsula, and Southern Ocean.
	Polar regions contain important drivers of climate change which, once triggered, may continue for centuries, long after greenhouse gas concentrations are stabilized. They may have irreversible effects on ice sheets, global ocean circulation, and sea-level rise.

Source: IPCC (2001).

detail and discussion can be found in Price and Barry (1997). Mountain resources that provide food and fuel for regional populations may be disrupted in developing countries. A general trend is that plant and animal species are expected to shift to higher elevations (Grabherr, Gottfried, and Pauli 1994; Gottfried, Pauli, and Grabherr 1998). Some species limited to mountain summits could become extinct. However, these changes are complex, as shown by studies in Japan (Nishioka and Harasawa 1997; Omasa et al. 2001). Snow accumulation plays a key role in causing differences in local vegetation (Yoshioka and Kanako 1963). The vertical distribution zone and domination degree of beech (*Fagus crenata*) differ between the Pacific Ocean and Japan Sea coasts of Japan, mainly because of the different patterns of snow accumulation resulting from the contrasting situations – lee and windward to the winter monsoon. Modelling by Tanaka and Taoda (1996) has shown that, on the Japan Sea coast, snow accumulation will be reduced, owing to the predicted weakening of the winter monsoon, resulting in shrinkage of the altitudinal zone and a decrease in the domination of beech forests. In contrast, on the Pacific coast, snow accumulation will increase, owing to an increasing frequency of cyclones. Together with the warming effect, this will result in a widening as well as a raising of the altitudinal zone and an increase in the domination of beech forests.

Tourism and recreation are likely to be disrupted, both directly and indirectly. For example, in the case of trekking and mountaineering, forest degradation (due to climatic change as well as to an increasing demand for fuelwood caused by greater numbers of visitors) would be affected by global warming. Melting mountain permafrost would lead to hazards, such as debris flows and rockfalls, and would change the distribution of fauna and flora, which are key tourism resources. Melting glaciers lead to GLOF (see below). Whereas there has been some work on the impacts of melting mountain permafrost in the Alps, little attention has been paid to melting hazards in the Himalaya, the Andes, and many other mountain regions (see also section on hazards, below). For winter sports, rising snowlines are a key concern: in Switzerland, a 2°C warming would bring a decrease in winter sports annual revenue of US$1.7 billion (Müller 1999). In the high mountains of Japan, Inoue and Yokoyama (1998) estimated that a similar warming would result in a decrease of 16 per cent in the proportion of precipitation falling as snow. Such factors as the melting of permafrost, changing fog (cloud) line, increasing levels of ultraviolet radiation, a decrease in frost days, and the increased use of snow cannons, would all affect winter sports. In addition to these direct impacts of global warming, new competition from other destinations would also affect economies based on winter sports in mountain areas.

Climate-change studies require climate data over a long period; however, climate data for mountain regions are not complete, and records do not usually extend over long periods. The Alps and parts of the Carpathians have the densest networks and longest records, extending back into the eighteenth century. Relatively dense networks also exist for the mountains of Britain, the Caucasus, Scandinavia, parts of North America, and the northern Andes (Barry 1992; Price and Barry 1997), but limited access and resources have limited the installation and efficiency of weather stations in other regions.

Regional haze

As well as the impacts of greenhouse gases, the effects of regional haze are also becoming an emerging challenge for some mountain regions. For example, the recent Indian Ocean Experiment (INDOEX) revealed a brownish haze layer over the Indian Ocean more than 1,000 km off the coast. Haze affects climate and environment in many different ways: observational results and climate-modelling studies (UNEP and C4 2002) suggest that the haze layer could have potentially significant impacts on monsoon climate, water stress, agricultural productivity, and human health. The most direct effects include a significant reduction in the amount of solar radiation reaching the surface, a 50–100 per cent increase in solar heating of the lower atmosphere, suppression of rainfall, reduction in agricultural productivity, and adverse health effects.

Aerosols can directly alter the hydrological cycle by suppressing evaporation and rainfall. With respect to agricultural production, decreases in the amount of solar radiation received by vegetation can impact productivity both directly and also indirectly through the induced changes in temperatures and hydrological cycle. Model simulations (UNEP and C4 2002) show that rainfall disruption is surprisingly great. This will be a concern both in mountain regions and downstream from them. Simulations also show compensated drying during the winter over areas northwest of India and over the west Pacific. These changes in precipitation are roughly consistent with recently observed trends. These studies represent very early stages of understanding the impact of haze on regional climates (UNEP and C4 2002).

Water

The Ministerial Declaration of the Second World Water Forum in the Hague, the Netherlands (March 2000) identified water security as a principal concern for sustainable development in this century. At the global scale, it is estimated that approximately one in three people live in regions of moderate-to-high water stress and that two-thirds of people will live in water-stressed conditions by 2025 (UNEP 1999; WBGU 1999).

Over 90 per cent of the earth's fresh water is stored as ice which, together with seasonally stored snow, provides melt flows into rivers during the hot, dry seasons. This is one of the reasons for mountains being described as "water towers" – the sources of fresh water for billions of people around the world – including about three billion people in China, South-East Asia, and South Asia, who depend on the rivers flowing from the Tibetan plateau. All of the world's major rivers originate in mountains: between one-third and one-half of all freshwater flows come from mountain areas; more than one-half of humanity relies on mountain water for drinking, domestic use, fisheries, irrigation, hydroelectricity, industry, recreation, and transportation.

Although mountain areas occupy only relatively small proportions of most river basins, they play a critical role in regional hydrological cycles, not only because their greater height triggers precipitation but also because temperature decreases with altitude; this means that there is less evaporation once the precipitation has fallen, and also that it is more likely to fall as snow than as water. For people living in the lowlands below, the storage of winter precipitation as snow or ice is especially crucial, because this melts when temperatures rise in the spring and summer. The water that is released enters the rivers, flowing downstream exactly at the time when it is most needed in the lowlands, sometimes thousands of kilometres away, for irrigation and other uses. This is most important in the dry parts of the world, where mountains are often the only areas receiving enough precipitation to generate runoff and recharge groundwater, typically providing 70–95 per cent of the flow to nearby lowlands. Even in humid areas, mountain water contributes 30–60 per cent of the water flowing to the lowlands. In Europe, although the Alps cover only 23 per cent of the area of the Rhine river basin, they provide one-half of the total flow. Other parts of the Alps form one-third of the area of both the Rhone and Po river basins and contribute 47 and 56 per cent, respectively, to the lowland flow (Mountain Agenda 1998). Consequently, mountain areas play a major role in determining the global water supply.

Owing to anthropogenic pressures, such as climate change, there have been major hydrological changes in mountain areas in recent years. Unpredicted river flows, and frequent floods, droughts, and crop failures are becoming more frequent. The management and protection of water resources have reached a critical period. The major challenges for mountain water resources include global climatic changes that are already beginning to affect water supply and demand, surface and groundwater contamination from pollutants, increased occurrence of water-related diseases, and the degradation of freshwater ecosystems.

A key issue is the loss of mountain water resources due to the shrinkage of glaciers, regarded by the IPCC (2001) as among the most unequivocal evidence for global climate change. For example, owing to a temperature increase of 1°C, the glaciers of the Alps have shrunk by 40 per cent in area and by more than 50 per cent in volume since 1850. In Africa, the glaciers of Mount Kilimanjaro are receding rapidly, with a decrease of 82 per cent in cover from 1912 to 2000. It is predicted that, by 2015, these glaciers will have disappeared (CSE 2002).

The Himalayan glaciers are also melting rapidly. These glaciers are extremely sensitive to global warming because they accumulate snow during the monsoon season and shed it in the summer. The melting of the glaciers is important, not only with regard to long-term water supplies but also because of the increased risk of GLOF. A recent study conducted by UNEP and the International Centre for Integrated Mountain Development (ICIMOD) identified 3,252 glaciers and 2,323 glacial lakes in Nepal and 677 glaciers and 2,674 glacial lakes in Bhutan (ICIMOD and UNEP 2001). On the basis of actively retreating glaciers and other criteria, the potentially dangerous glacial lakes were identified using the spatial and attribute database complemented by multi-temporal remote sensing and evaluation of the active glaciers. The study also confirmed that groups of closely spaced supraglacial lakes of smaller size at glacier tongues merge over time, forming larger lakes; these are indications that lakes are growing rapidly and becoming potentially dangerous.

Glaciers in other parts of the Himalaya have yet to be studied and documented with a similar methodology to that used in the Bhutan and

Nepal study. Such work is essential for the development of early warning systems for the Hindu Kush–Himalayan region. The problem is likely to be widespread in other regions with glaciers: experience from the Alps has shown that even minor GLOF can have catastrophic consequences. Bursting of glacial lakes and fast glacier-recession rates would cause large-scale flooding and mudslides and eventual drying-up of the rivers. This would have important consequences for water supplies, hydroelectricity generation, riparian habitats, and tourism, and could lead to more frequent drought, crop failure, and poverty.

Hazards

Many hazards are associated with mountain-building and mountain environmental processes. These hazards are mainly in the form of earth-surface processes, such as snow avalanches, rockfalls, debris flows, volcanic mudflows (lahars), glacial lake outburst, and other types of floods. These processes are influenced by relief (steepness of slopes, ruggedness of topography), lithology, landform history, and precipitation events. Some natural hazards, such as snow avalanches and catastrophic rockslides, occur only (or largely) in mountain areas. Others, such as earthquakes, debris flows, and volcanic eruptions, are more common or more severe in mountain areas. However, most types of hazards found in mountain areas – for example floods, droughts, and forest fires – also occur in other regions. Hazards are major environmental constraints in sustainable development in mountain areas.

The Disaster Database of OFDA/CRED (1991–2000) records that a total of 2,557 hazards were reported from 1991 to 2000, worldwide. These include avalanches/landslides, droughts/famines, earthquakes, extreme temperature, floods, forest/scrub fires, volcanic eruptions, windstorms, and other natural hazards. Of this total of 2,557 disasters, 173 were avalanches/landslides, 223 drought/famines, 221 earthquakes, 112 extreme temperature, 888 floods, 123 forest/scrub fires, 55 volcanic eruptions, 748 windstorms, and 25 other natural disasters. The data show that 665,598 people were killed (fig. 2.4) and that the total damage amounted to US$692.9 billion (fig. 2.5).

Figures 2.2 and 2.3 clearly illustrate that Asia is more vulnerable than the other four continents in terms of people killed and damage caused. The mountains of Asia (particularly of South Asia) are characterized by high relief, very intense tectonic activity, highly concentrated precipitation, and high population density, all of which make these regions susceptible to natural hazards and disasters. The major triggering factors for

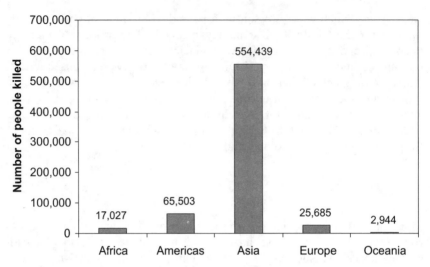

Figure 2.2 Total number of people reported killed by various natural disasters, by continent. Data from Disaster Database of OFDA/CRED (1991–2000)

landslide and debris flows are heavy rainstorms, snowmelt runoff, earthquakes, volcanic activities, and human modification of mountain slopes.

Natural dams created by landslide and avalanches are also a significant

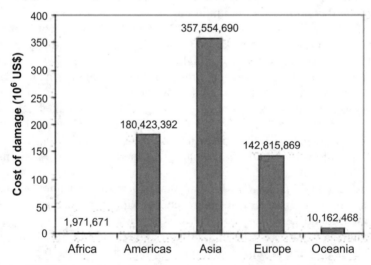

Figure 2.3 Total amount of estimated damage caused by various natural disasters, by continent. Data from Disaster Database of OFDA/CRED (1991–2000)

hazard in mountain areas, and are particularly common in the high rugged Hindu Kush–Himalaya in South Asia and the Hengduan Mountains in south-west China. Casualties from individual landslide dam failures have reached many thousands. The world's worst recorded landslide dam disaster occurred when the 1786 Kangding-louding earthquake in Sichuan Province, China, triggered a huge landslide that dammed the Dadu River: after 10 days, the landslide dam was overtopped and breached; the resulting flood extended 1,400 km downstream and drowned about 100,000 people (Tianchi, Schuster, and Jishan 1986).

More recently, the Yigong River in south-eastern Tibet, China was dammed on 9 April 2000 by a huge landslide. After two months, the dam partially failed on 10 June 2000. A flash-flood more than 50 m high travelled more than 500 km downstream of the landslide dam site. This very high-speed flood damaged many bridges and 70 km of highway, created numerous new landslides along both sides of the river, and changed the landscape and hydrological regimes in many sections of the Yigon, Palong, and Brahmaputra rivers. The flash flooding also resulted in 30 deaths, more than 100 people missing, and more than 50,000 homeless in the five districts of Arunanchal Pradesh, India (Tianchi, Zhu, and Yongbo 2001).

Losses from natural hazards in mountain areas have been increasing as the result of such factors as overexploitation of natural resources and deforestation, and the construction of infrastructure such as buildings, roads, irrigation canals, and dams. This trend is likely to be magnified by changes in precipitation regimes and increases in extreme events likely to result from global climate change. For example, on the Tibetan plateau, it has been predicted that 5 per cent of the permafrost in the high mountain areas will melt in coming decades and that landslides and debris flows will become more severe in high mountain areas (Chen 1996).

Desertification

The formal definition of desertification adopted by the United Nations Convention on Desertification is "land degradation in arid, semi-arid, and dry sub-humid areas resulting from various factors, including climatic variations and human activities." Inclusion of climatic variation in the definition itself shows the influence of climate change in desertification. In general terms, desertification refers to the reduced ability of land to support vegetation, leading to a vicious cycle of poor vegetation and poor soil.

Despite the fact that desertification has become a global issue, it remains poorly understood. Available estimates of areas affected range from one-third to about one-half of the world's land area, and people affected from 1 in 6 to 1 in 3 (Toulmin 2001). One common estimate is that desertification/land degradation affects almost 30 per cent of the global land area and nearly 850 million people. The problem of desertification has been becoming more and more urgent each year: for example, the deserts of China are expanding each year by 2,460 km^2, at a cost of US$6.52 billion (Reuters, 21 March 2002).

Desertification is caused by complex interactions among physical, biological, political, social, cultural, and economic factors. Factors encouraging degradation in mountain areas include climatic variation and unsustainable human activities such as overcultivation, overgrazing, deforestation, or poor irrigation practices. The main unfavourable social, cultural, and political factors include low literacy rates, high female workloads, and lowland interests. In Africa and in North and South America, "very degraded" soils are mostly found in mountain areas.

Deserts are likely to become hotter but not significantly wetter with the impacts of climate change on hydrological systems. With the reduction of flows from mountains in the dry season, deserts may well expand into mountain areas. Warmer conditions could threaten desert species living near the limit of their heat tolerance. Desertification is more likely to become irreversible under drier conditions and when land has been further degraded through erosion by high-intensity precipitation.

Implications: Best practices

Mountain issues cannot be separated from issues and activities in the lowlands, especially in the context of emerging atmospheric issues. These issues will pose major challenges for mountain areas and their natural resources in the foreseeable future. In this section the best practices for policy development and practical implementation are suggested and existing initiatives and partnerships are analysed.

Policy development and implementation

Mountain issues cannot be tackled by mountain communities or by individual countries alone. This is particularly true for atmospheric issues, which derive from regional to global processes. Consequently, partnerships between institutions and programmes concerned with mountain and atmospheric issues are vital, so that the issues can be tackled jointly. Therefore, *regional agreements* should be adopted, recognizing the need for joint action. Such agreements should not limit their scope only to political dialogue: under the framework of such agreements, national policies should be developed to establish the *scientific base* for understanding these issues.

The development and distribution of relevant educational material and information on climate change and its implications on mountain environments and socio-economic consequences are also vital in order to move the policy cycle forwards. The existing conservative approach to data sharing should be changed and *dissemination of scientific findings* should be encouraged.

Practical implementation

In order to cope with hazards such as GLOF, *early warning systems* should be developed and implemented using a multi-stage approach, multi-temporal data sets, and multi-disciplinary professionals. The initial focus should be on known, potentially dangerous, "hot spots." The development and implementation of monitoring, mitigation, and early warning systems involve several phases. The following list, adapted from ICIMOD and UNEP (2001), shows possible steps for GLOF monitoring and mitigation and for early warning systems and their implementation in Nepal:

- Detailed inventory and development of a spatial and attribute digital database of the glaciers and glacial lakes using reliable medium- to large-scale (1:63,360 to 1:10,000) topographic maps.
- Updating of the inventory of glaciers and glacial lakes and identification of potentially dangerous lakes using remote-sensing data, e.g. the Land Observation Satellite (LANDSAT) Thematic Mapper (TM), Indian Remote Sensing Satellite (IRS)1C/D Linear Imaging and Self Scanning Sensor (LISS)3, Système Probatoire d'Observation de la Terre (SPOT) multi-spectral (XS), SPOT panchromatic (PAN) (stereo), and IRS1C/D PAN (stereo).
- Semi-detailed to detailed study of the glacial lakes, identification of potentially dangerous lakes, and the possible mechanism of GLOF using aerial photography.
- Annual examination of medium- to high-resolution satellite images (e.g. those from LANDSAT TM, IRS1D, SPOT, etc.) to assess changes in the different parameters of potentially dangerous lakes and the surrounding terrain.
- Brief over-flight reconnaissance with small-format cameras to view the lakes of concern more closely and to assess their potential for bursting in the near future.

- Field reconnaissance to establish clearly the potential for bursting and to evaluate the need for preventative action.
- Detailed studies of the potentially dangerous lakes by multi-disciplinary professionals.
- Implementation of appropriate mitigation measure(s) in the potentially highly dangerous lakes.
- Regular monitoring of the site during and after the appropriate mitigation measure(s) have been carried out.
- Development of a telecommunications and radio broadcasting system integrated with on-site installed hydrometeorological, geophysical, and other necessary instruments at lakes of concern and downstream as early warning mechanisms for minimizing the impact of GLOF.

Early warning systems should be supported by continuous *monitoring* of key environmental variables. This requires the establishment of, and long-term support for, observatories for air quality, and for meteorological and aerosol monitoring. Hot-spot areas should be given priority when establishing the observatories. Together with satellite observations, data from these observatories should provide critical coverage for the understanding of long-term trends.

A more complete picture of the roles and interactions of greenhouse gases, aerosols, and ozone is urgently needed. The aerosols and high-level ozone that result from rural and urban air pollution are implicated in global warming, because they could influence climate change by altering radiative balance on a regional, and perhaps global, scale. Their presence can also have effects on the ecosystem, particularly on vegetation. Thus, there is a need to assess impacts within a coherent framework. For this reason, not only monitoring but also coordinated *scientific studies* complementing observatory results should be conducted.

Research initiatives

Geo-ecological studies in high mountains have been carried out since the mid-twentieth century. The "Field guide for landscape ecological studies in high mountain environments" (Winiger and Bendix 2000) proposes standards for field observations and instrumented networks in mountain areas. It focuses on selected basic ecosystem components, such as climate, geomorphology, soil, and vegetation. The potential use of vegetation as an environmental indicator, as well as the research potential of remote-sensing and geographical information system (GIS) techniques and model approaches, have been discussed for high-mountain research. There is also a long tradition of mountain climatology and meteorology

(Yoshino 1975; Barry 1992). The World Climate Research Programme (WCRP) implemented ALPEX (Alpine Meteorology Experiment) as an international programme, which resulted in much new information.

Bringing a diverse range of approaches and initiatives from different disciplines together, the Mountain Research Initiative (MRI; Becker and Bugmann 2001) has been developed within the context of the International Geosphere–Biosphere Programme (IGBP), the International Human Dimensions Programme on Global Environmental Change (IHDP), and the Global Terrestrial Observing System (GTOS). It includes four activities:

1. Long-term monitoring and analysis of indicators of environmental change in mountain regions;
2. Integrated model-based studies of environmental changes in different mountain regions;
3. Process studies along altitudinal gradients and in associated headwater basins;
4. Sustainable land use and natural-resource management.

Some of the elements of these activities are quite well developed, for example for monitoring glaciers (Haeberli, Barry, and Cihlar 2000) and alpine plants (Pauli et al. 2001) within activity 1.

Institutional initiatives

Most national governments have established national institutions for sustainable development. Mountain issues are part and parcel of the national environmental issues and are addressed by such institutions. In addition, there is increasing coordination of mountain initiatives between countries under transboundary provisions because (although many mountain ranges are divided by national boundaries) their utilities and management involve cross-national links. A good example is the International Centre for Integrated Mountain Development (ICIMOD) in the Hindu Kush–Himalayan region, inaugurated in December 1983 with a coordinating role in this region.

Although national and international efforts are essential to improve the sustainable management of natural resources in mountain areas, it is also necessary to tackle the emerging atmospheric challenges. Because these are transboundary in nature, they can be addressed only through intergovernmental cooperation. The Convention on Long-range Transboundary Air Pollution (for Europe), the Malé Declaration on Control and Prevention of Air Pollution and Its Likely Transboundary Effects for South Asia (for South Asia), and the East Asian Network on Acid Depositions (EANET) (for East Asia) are good examples of regional cooperation in tackling such issues.

At the international level, a vigorous response to climate change – involving research, discussions, planning, and implementation – started in 1988 with the establishment of the Intergovernmental Panel on Climate Change (IPCC) by UNEP and WHO (the World Health Organization). This has resulted in the 1992 Convention on Climate Change and the 1997 Kyoto Protocol. This latter incorporates legally binding targets for the reductions in emissions of greenhouse gases. In order to meet these targets, a number of flexible mechanisms have also been developed.

Key actions

National governments and institutions

- Develop a systematic and continuous *monitoring* system for monitoring mountain environments. The system should cover the three major components of mountain environments – land, air, and water.
- Raise awareness and provide *early warning* information with respect to changes in mountain environments and their consequences. The target groups should not be limited only to the mountain communities: the messages should also reach lowland communities.
- Support regional and international *research* initiatives into the various elements of mountain environments, with particular emphasis on their interactions and influences on human societies.
- Make full use of existing *conventions*.

International institutions and donors

- Document *available technologies and best practices*, whether modern or traditional.
- Disseminate an *inventory of mitigation options and best technologies* to national institutions and mountain communities.
- Ensure *capacity building* of national institutions for monitoring mountain environmental issues in developing countries. This should include continual monitoring, complemented by research projects and programmes.
- Build *partnerships* linking the several international conventions and agreements calling for sustainable management of land and water resources. These objectives are often potentially affected by climate change. To the extent possible, options to adapt to changing climatic conditions can be structured to help attain environmental and socio-economic objectives associated with these other agreements.

The status and the challenges for mountain environments will change, but the momentum initiated by the International Year of Mountains should be continued. It is proposed that a biennial assessment of the status of mountain environments should be implemented and published, with a definition of the challenges and proposals for meeting them.

REFERENCES

Barry, R.G. 1992. *Mountain weather and climate*. London: Routledge.

Becker, A., and H. Bugmann (eds). 2001. *Global change and mountain regions: The Mountain Research Initiative*. Stockholm: IGBP Secretariat, Royal Swedish Academy of Sciences.

Chen, Bangin. 1996. "Possible impacts of global warming on natural disasters." *Journal of Natural Disasters* Vol. 5 No. 2, pp. 95–101.

CSE. 2002. *Down to Earth*, 15 May 2002. New Delhi: Centre for Science and Environment.

Diaz, H.F., and R.S. Bradley. 1997. "Temperature variations during the last century at high elevation sites." *Climatic Change* Vol. 36, pp. 253–279.

Du, Jun. 2001. "Change of temperature in Tibetan Plateau from 1961 to 2000." *Acta Geographica Sinica* Vol. 56, No. 6, pp. 682–690.

Gottfried, M., H. Pauli, and G. Grabherr. 1998. "Prediction of vegetation patterns at the limits of plant life: A new view of the alpine–nival ecotone." *Arctic and Alpine Research* Vol. 30, No. 3, pp. 207–221.

Grabherr, G., M. Gottfried, and H. Pauli. 1994. "Climate effects on mountain plants." *Nature* Vol. 369, pp. 448–449.

Haeberli, W., R.G. Barry, and J. Cihlar. 2000. "Glacier monitoring within the Global Climate Observing System." *Annals of Glaciology* Vol. 31, pp. 241–246.

ICIMOD and UNEP. 2001. *Inventory of glacial lakes and glacial lake outburst floods: Monitoring and early warning systems in the Hindu Kush–Himalayan Region*. Kathmandu: ICIMOD.

Inoue, S., and H. Yokoyama. 1998. "Estimation for vegetation of snowfall and accumulation under the global warming." *Seppiyo* Vol. 60, No. 5, pp. 367–378 (in Japanese).

IPCC. 2001. "IPCC third assessment report: Climate change 2001. Working Group II: Impacts, adaptation and vulnerability. Summary for Policy Makers." Geneva: WMO and UNEP.

Mountain Agenda. 1998. *Mountains of the world: Water towers for the 21st century*. Berne: Mountain Agenda.

Müller, H. 1999. "Tourism and climate change." In: *Mountain Agenda, Mountains of the world: Tourism and sustainable mountain development*. Berne: Mountain Agenda.

Nishioka, S., and H. Harasawa. 1997. *Global warming and Japan*. Tokyo: Kokonshoin (in Japanese).

Omasa, K. et al. 2001. "Impacts on terrestrial ecosystems." In: *Impacts of global warming on Japan 2001*. Working Group on Impact of Global Warming and Its Assessment, Committee on Global Warming Problems. Tokyo: Ministry of the Environment (in Japanese).

Office of U.S. Foreign Disaster Assistance (OFDA) and Centre for Research on the Epidemiology of Disaster (CRED): OFDA/CRED International Disaster Database, http://www.who.dk/ccashh/extreme/1991–2000.

Pauli, H., M. Gottfried, K. Reiter, and G. Grabherr. 2001. "High mountain summits as sensitive indicators of climate change effects on vegetation patterns: the 'Multi Summit-Approach' of GLORIA (Global Observation Research Initiative in Alpine Environments)." In: G. Visconti et al. (eds) *Global Change and Protected Areas*. London: Kluwer.

Price, Martin F., and R.G. Barry. "Climate change." In: B. Messerli and J.D. Ives (eds) *Mountains of the world: A global priority*. Carnforth: Parthenon.

Reuters, 21 March 2002.

Shrestha, A.B. et al. 1999. "Maximum temperature trends in the Himalaya and its vicinity: An analysis based on temperature records from Nepal for the period 1971–94." *Journal of Climate* Vol. 12, pp. 2775–2786.

Tanaka, N., and H. Taoda. 1996. "Expansion of elevational distribution of beech (*Fagus creneta* Blume) along the climatic gradient from the Pacific Ocean to the Sea of Japan in Honshu, Japan." In: K. Omasa et al. (eds) *Climate change and plants in East Asia*. Tokyo: Springer.

Tianchi, Li, R.L. Schuster, and Wu Jishan. 1986. "Landslide dams in South-Central China." In: R.L. Schuster (ed.) *Landslide dams: Processes, risk and mitigation*. American Society of Civil Engineers, Special Publication 3.

Tianchi, Li, Pingyi Zhu, and Chen Yongbo. 2001. "Natural dam created by rapid landslide and flash flooding from the dam failure in Southeastern Tibet, China, 2000." Paper presented in the Regional Workshop on Water-Induced Disasters in the Hindu Kush Himalaya Region, Kathmandu, Nepal, 2001, December 11–14, unpublished.

Toulmin, C. 2001. "Lessons from the theatre: should this be the final curtain call for the convention to combat desertification?" International Institute for Environment and Development, WSSD opinion series.

UNEP. 1999. *Global environment outlook 2000*. London: Earthscan.

UNEP and C4. 2002. "The Asian brown clouds: Climate and other environmental impacts." A UNEP Assessment Report based on the Findings from the Indian Ocean Experiment.

Van Aardenne, J.A. et al. 2001. "A 1 degrees × 1 degrees resolution data set of historical anthropogenic trace gas emissions for the period 1890–1990." *Global Biogeochemical Cycles* Vol. 15, No. 4.

WBGU (Wissenschaftlicher Beirat der Bundesregierung Globale Umweltveränderung). 1999. "World in transition. Ways toward sustainable management of freshwater resources." German Advisory Council on Global Change, 1997 Annual Report.

Winiger, M., and J. Bendix. 2000. "Field guide for landscape ecological studies in high mountain environments." Draft prepared by members of the Working

Group on Mountain Geoecology, VGDH, paper presented at Comm. 96, C16, 29th International Geographical Congress, Seoul.

Yoshino, M. 1975. *Climate in a small area.* Tokyo: University of Tokyo Press.

Yoshino, M. 2002. "Global warming and mountain environment." Paper presented at the International Symposium on Conservation of Mountain Ecosystems, 1 February 2002, United Nations University, Tokyo.

Yoshioka, K., and T. Kanako. 1963. Distribution of plant communities on Mt. Hakkoda in relation to topography. *Ecological Review* Vol. 16, No. 1, pp. 71–81.

3

Mountain infrastructure: Access, communications, and energy

Thomas Kohler, Hans Hurni, Urs Wiesmann, and Andreas Kläy

Summary

Access, communications, and energy are very powerful agents of change, not only (but especially) in mountain areas. They involve vital linkages between mountain regions and adjacent downstream areas with their centres of population and economic activities. Issues in communications on the negative side relate to the loss of skilled personnel ("brain drain"), overexploitation of resources, environmental pollution, and disruption of local livelihoods; on the positive side they relate to employment generation, better access to health and education, and exposure to the wider world, including markets. Energy issues include the future role of hydropower, especially large-scale initiatives, the still-increasing use of fuelwood and other biofuels, and the potential and affordability of modern energy alternatives such as solar, wind, and passive solar energy.

In recent decades, extensive road construction has improved access and communication in many mountain areas – a trend that is likely to continue. Specific technologies such as ropeways, suspension bridges, or air transport can provide links where roads and railways are not economic. Modern communications technologies have shown their great potential in marketing (tourism, sale of mountain products), telemedicine, and distant education. Mountains have extensive potential for hydropower generation, but this potential is still largely untapped in developing countries. Fuelwood is the main source of energy in many mountain areas and is likely to retain this position for many years to come. Unfortunately, many modern alternatives such as solar and wind power are not mass energy producers, but have potential in mountain areas as stand-alone facilities in remote places. A potential yet to be fully recognized is passive solar use, including insulation of buildings.

The development of transport, communications, and energy should adhere to four basic principles: negotiated outcomes, shared benefits, tailored (technical) solutions for mountains, and building on existing facilities and experiences. Keywords for best practices are decentralized, small-scale, and phased development. This is true especially for road access and hydropower development, which involve higher costs for construction and maintenance than in downstream areas, owing to difficult terrain, environmental risks, and natural hazards. Where mountain resources such as hydropower or timber are exploited for downstream interests, adequate compensation must be made to mountain communities. Credit schemes will be important to promote the development of access, communications, and energy in mountain areas. Although traditional modes of transport and energy must not be forgotten, increasing the efficiency of use should be a priority. Animal transport often remains important and should be supported by specific research and breeding programmes and adequate veterinary services.

The development of access, communications, and energy involves a concerted effort of key stakeholders. These include local communities, national and regional governments, civil society and NGOs (both national and international), international development organizations and the donor community, the private sector and professional associations, and the scientific and research community.

The issues

Access, communications, and energy are key issues in the sustainable development of mountain areas. Experience has shown that they are very powerful agents of change, not only (but especially) in mountains. Access, communications, and energy in mountain regions also involve vital linkages between these regions and adjacent lowlands, centres of population, and industrialized and urbanized areas.

Issues relating to access and communication

Despite their relative isolation, mountains have always depended to various extents on exchange and trade with surrounding areas. This is evident from the existence of long-established markets in mountains, including areas dominated by subsistence mountain agriculture. Likewise, there have been transit routes across mountains since ancient times. With the advent of industrialization, the need to move goods in large quantities, and the development of modern means of transportation such as railways and roads, mountains have increasingly been drawn into networks of access, transit, and communication. The issues related to this development centre around the impacts of this increased accessibility, which is a global phenomenon, and whether they have a positive or negative impact on sustainable mountain development.

Key negative factors in the debate include "brain drain" (the loss of skilled personnel), overexploitation of resources, disruption of local livelihoods, and labour migration; employment, better access to health and education, and exposure to the wider world are positive aspects for mountain development. Interestingly, the appearance of modern communication technologies (Internet, email) in mountain areas is an issue that is much discussed in development circles. The focus on modernity and modernization should not make us forget that traditional means, such as animal power, are still the most important form of transport (alongside human power) for most mountain people, especially in developing countries. However, ways and means of improving this form of transport are not on the development agenda and elicit very little in terms of research and development funds.

Energy issues

Mountains are vital sources of energy in very diverse forms. Specifically, mountains and highlands are known for their extensive potential for hydropower generation, which is due to high gradients, relatively high precipitation and runoff compared with adjacent lowlands, and water stored as snow and ice. More electricity is produced downstream on rivers fed by mountain areas than in other locations. Large-scale versus small-scale hydropower development, and compensation of mountain communities for the use of water resources for hydropower generation (which is mainly used to serve downstream interests), are key issues in the debate over energy production. The potential of hydropower to help move the world away from its present high level of consumption of fossil fuels, which is an important factor in global warming, adds a new dimension to the hydropower debate. For most mountain people, however, fuelwood is still the most important source of energy. The demand is increasing as populations grow. Key issues include the efficiency of fuelwood use and the search for viable and proven alternative energy sources – such as solar and wind power, biogas, and fuel cells.

Knowledge

Access and communication

Mountains have long been seen as obstacles to movement, even though people have developed routes into mountains and across them since ancient times. Ancient transit routes include the Silk Road linking China to Europe across the mountains of Central Asia, routes built by the Roman Empire across the European Alps, and routes in the Andes established by the Incas. Old marketplaces and market villages also show that mountains have always been in need of exchange with the outside world, even in areas where subsistence agriculture has dominated local economies. This illustrates the fact that exchange and, hence, access, is – and always has been – part and parcel of the concept of subsistence, even in seemingly isolated and secluded mountain areas.

Industrialization and mass mobility on a global scale have greatly increased accessibility and communications networks. In mountain areas, access and transit infrastructure have been improved mainly by extensive road construction in recent decades in many parts of the world. However, density of access still differs greatly between mountain regions in industrialized and developing countries. Switzerland's road network, for example, is nearly 100 times denser per unit area, and 23 times denser per capita, than that of Ethiopia. The 100 per cent accessibility in Switzerland means that every household in a given area reaches its home directly by car; in Ethiopia, it means that every household can reach the next motorable road within a day's walking time (Schaffner and Schaffner 2001).

Mountains have benefited from transport development through increased employment and income opportunities, notably through daily or seasonal labour migration to surrounding lowland areas, which has been greatly facilitated especially by road connection. Improved access has facilitated the development of local markets, small and large industries, and services such as tourism. All of this has created much-needed local employment and increased economic diversification. Road connection has helped to stabilize population numbers and the quality of life in many places, as it has made possible the development of basic infrastructure such as clinics and schools. Other benefits have included access to health, education, and consumer goods, and exposure to the wider world. External support to mitigate the effects of natural disaster or famine can be much more effectively provided with adequate access, for instance in Ethiopia. Access has also increased opportunities for regional cooperation and economic exchange between mountain areas. In the European

Alps, for example, 60 per cent of the total traffic volume (i.e. distances travelled in kilometres) is local and regional traffic within the Alps. This stands in direct contrast to the public debate, which focuses almost exclusively on the issue of transit traffic in the countries concerned.

On the other hand, increased accessibility by means of mass transportation – railways, roads, and ropeways – has resulted in a number of negative impacts which include "brain drain" and overexploitation of resources, for example in mining. On the eastern slopes of the Peruvian Andes, for example, roads have facilitated the rush for gold and timber since the early 1990s. This boom was made possible by importing over 1,000 pieces of heavy earth-moving machinery, which have caused extensive damage to the rain forests. The completion of the Transoceanic Highway between Peru and Brazil across the Peruvian Andes is likely to increase present levels of overexploitation and destruction of forests (Seimon 2001). Other examples of overexploitation include tourism and forest use: along the Karakorum Highway in Northern Pakistan, for example, forest-stand density of highly accessible forests (closest jeepable road within 2 km) has decreased by up to 85 per cent as opposed to 0–40 per cent in less-accessible forests (closest jeepable road further than 8 km) (Schickhoff 2001). On the other hand, the Highway is the lifeline for Northern Pakistan with its ever-increasing food deficit (see Case 1).

Case 1. Karakorum Highway: Lifeline for Northern Pakistan

As a result of the Pakistan–China Border Treaty of 1963, bilateral cooperation led to the construction of the Pakistan–China Friendship Highway, commonly referred to as the Karakorum Highway (KKH).

In addition to its obvious military importance, the KKH has become the lifeline for the mountains of Northern Pakistan. Cereals and fresh meat, imported as live animals for slaughter in local bazaars, account for more than three-quarters of all goods from the lowlands, which supply army personnel, tourists, and a growing number of local farmers and traders. It has reduced prices for lowland goods such as chemical fertilizer, which previously cost twelve times as much in Gilgit when it was airlifted. As a result of subsidized food imports from Pakistan's lowlands, the proportion of food produced locally is steadily decreasing: in some villages of the Hunza Valley, local production is now less than one-third of the household's annual consumption. The KKH has been instrumental in mitigating food shortages and, for the first time in history, there are now no periods of starvation and famine, as a result of crisis management by the Federal Government and the World Food Programme. Development has followed suit, greatly improving general infrastructure, education, and health services, and facilitating the introduction of niche products such as seed potatoes, vegetable seeds, and fruits. It has also led to the construction of an extensive network of secondary roads linked to the KKH.

However, possible drawbacks should be kept in mind. Closure of the KKH due to natural hazards or human intervention could have severe results. Although the KKH engineer corps are maintaining the road and deal effectively with natural hazards, especially in spring and during the monsoon season, control is much more difficult when highway robbers or politically motivated activists threaten the safety of the KKH, making use of its crucial role to suit their interests. Source: MF-Asia Moderator (2002), taken from Kreutzmann (2000): Improving accessibility for mountain development.

Better access has led to the destruction or fragmentation of mountain habitats, including forests, and disruption of local culture in many parts of the world's mountains. However, where access increases the availability and reduces the cost of alternative energy sources such as gas and kerosene, it can help to reduce pressure on forest resources – as in the European Alps in the past. Where road traffic is heavy, as around towns and conurbations or in important transit corridors, it has led to high levels of air pollution, which have a negative impact on the local quality of life and the environment. Examples of such problems include the European Alps, Southern North America (Mexico City), and the Andes (Santiago de Chile).

The development of modern transport infrastructure, especially of roads and railways, is a costly enterprise. Costs in mountains are even higher than those in lowlands, for both construction and maintenance, owing to difficult topography; harsh climate; and the need both for protection from hazards such as avalanches, landslides, and rockfalls, and to secure road- and railside slopes. In Switzerland, the construction of one metre of a dual-track railway bridge can cost as much as US$40,000–80,000. Although such costs are prohibitive for many economies, labour-intensive road construction has shown great potential in many developing countries. Costs are lower than when heavy machinery is used, and the economic benefits of construction are largely retained within mountains. In Ethiopia, for example, 30–40 per cent of construction costs has gone directly to local people in the form of wages. There is also less environmental damage than when construction is based on the use of heavy machinery (Hartmann 2001; Schaffner and Schaffner 2001).

Specific technologies – such as ropeways, suspension bridges, or air transport – can provide links where railways and roads are not economic. In Nepal, over 1,000 suspension and suspended bridges have been constructed in recent years, providing access to hitherto secluded valleys and settlements (Gaehwiler and Lamichaney 2001). Ropeways, of which over 10,000 are in operation (mostly in industrialized countries and mainly in tourist areas), have considerable potential for hauling goods and people in mountains (Schmoll and Seddon 2001). Typically, construction and maintenance costs are lower than those for roads, especially for gravity ropeways. These installations are environmentally friendly and less susceptible to hazards, and local communities have greater economic control over trade and transport than they do with roads.

Modern communications technologies such as the Internet have been increasingly important to link institutions and individuals interested in mountain development on global, regional, and national levels. However, linkage and exchange are still affected by the "digital divide": people and institutions in mountains are still largely excluded from access to and use of these technologies, especially in developing countries. This is due to lack of the requisite infrastructure (telephone connections) and the high initial costs for the purchase of necessary equipment such as personal computers. In many mountain countries, access to the Internet is confined to the capital and a few larger towns, and the number of lines is very low indeed: for example, there are more telephone connections in Manhattan, New York, than in the whole of Africa.

Efforts should be directed towards reducing the digital divide, as modern communications technologies have proven their great potential (also for mountain regions) in such diverse applications as telemedicine, distance education, tourism promotion, and marketing of local products. Farms selling salmon in the remote Western Highlands of Scotland, for example, have reported a 30 per cent increase in sales following the use of the Internet (Price and Houston 2001). Many people in mountain areas, and in the mountain regions of developing countries, such as Nepal, have heard about these technologies. Simple telephone access has the most impact, as users do not have to be literate and no specific language skills (e.g. the use of English) are required. Interestingly, modern communications technologies could reverse the progression of access as we have witnessed it (starting with road access, followed by electricity supply, and finally telephone): today, telephones are often the first means of access to an area via radio call or satellites, followed by electricity and, lastly, roads, which are the most costly form of access (L.L. Montgomery, Internet discussion).

The focus on modern means of communication and transport, especially on road development, has resulted in neglect of traditional forms of movement such as animal transport. Animal transport is still the most important means for moving goods and people in many mountain areas of the world, and is very effective in difficult terrain. However, government officials and development experts are largely unaware of its potential and of key issues, as the topic is omitted from their training. Animal power is a natural renewable energy source: animals are integrated into local subsistence production systems, consume local feed, reproduce, supply valuable manure, and minimize environmental damage. They are widely available in mountains and generally affordable by local people. Animal transport is labour intensive and provides valuable employment to many. In the mountains of Ethiopia, five million donkeys carry water, fuelwood, and other merchandise to remote villages. One-third of the donkeys in the world are found in the Asian mountains, including the mountains of China and Pakistan (Starkey 2001). Yaks and cross-breeds between yaks and cattle are important pack animals for local transport and trekking tourism in the Himalaya. Llamas carry small loads in the Andes, for local and tourist demands (see Case 2), and camels are widely used for transportation in the mountains of the Middle East, Central Asia, and the Arab world.

Case 2. Tourism and animal transport: The reintroduction of the Hutsul horse, Bieszczady National Park, Carpathian mountains, Poland

The Hutsul horse originates from the Eastern Carpathians; therefore, Bieszczady National Park (BNP) in Southern Poland was the natural place to reintroduce this almost extinct breed, which combines local, Mongol, Turkish, and Arab heritage. It is adapted to mountain environments and finds its way easily in difficult terrain; its docile character makes it suitable for both mountain horseback riding (even for young visitors) and for therapy for disabled children. In 1993, when no more than 230 individuals of this breed remained in Poland, the Hutsul horse reintroduction project was launched on a former state-owned sheep-raising farm, acquired by the park authorities. The project was funded by the National Fund for Environment Protection and Water Management, using funds derived from fees for natural resources use and fines for polluting the environment by industry.

Hutsul horses are now used in many ways. BNP rangers use them for patrolling the park. Horses are used for transporting litter collected along tourist trails, for therapies for disabled children, and for horseback tourism organized by BNP or by local entrepreneurs. BNP offers training to inexperienced tourists and subsidizes training for local inhabitants who wish to take up horseback-riding activities. Horseback trails now total 142 km in length, allowing several day-long expeditions in the Polish section of the Park. In future, the trail is planned to cover the Slovak and Ukrainian Carpathians within the Carpathian Euroregion, thus allowing the Hutsul horse to come back to its cradle in the Czarnohora (Hutsul name for "Black Mountain") range. Source: Zbigniew Niewiadomski, email ⟨zbig-niew@wp.pl⟩ Internet discussion.

Energy

Hydropower provides more than 97 per cent of all electricity generated worldwide by new renewable sources (solar, wind, geothermal, biomass); a substantial portion of this is generated in mountains. As awareness of the need to move towards environmentally clean energy is increasing globally among politicians and decision makers, hydropower generation in mountains is likely to increase and its large potential, especially in developing countries, will increasingly be tapped. There is extensive experience worldwide with this mature technology, both with regard to the technology used and the institutional arrangements required for its installation, running, and maintenance.

Hydropower development differs greatly between developed and developing countries: whereas Norway has tapped over 65 per cent of its potential, countries such as Nepal and Ethiopia have developed less than 1 per cent of theirs (Pandey 2001). Many developing countries have recognized the potential of their mountain areas and have prepared ambitious plans for hydropower generation, including large-scale dams, of which the Three Gorges project in China is the most dramatic example. However, recent studies (World Commission on Dams 2000) have shown that large dams and hydropower schemes are often characterized by cost overruns, disruption of livelihoods, and destruction of habitats. They rarely benefit the mountain communities in territories where they are located, but are built mainly to serve the needs of downstream industrial and population centres (see Case 3). This is also true for industrialized mountain areas, such as the mountains in the Grisons in Eastern Switzerland, which exported 77 per cent of the electricity generated to downstream areas in the late 1990s.

Case 3. Voices carrying no weight: Mountain communities and major hydropower projects

The Lesotho Highlands Water Project (LHWP) is a bi-national, multi-purpose undertaking between the Kingdom of Lesotho and the Republic of South Africa (RSA). It is one of the world's largest infrastructure projects currently under construction, comprising five proposed dams and a 72 MW hydropower plant that will supply power to Lesotho. The first dam has delivered water since 1998 and the second is currently in the final phase of construction. Donors and lenders include the World Bank and an international consortium of other lending institutions, including commercial banks.

The project is subject to the usual debate about pros and cons of large-scale hydropower development. Often lost in this debate are the voices of the local people themselves. In the case of the LHWP, however, the Panos Oral Testimony Programme (Mountain Voices: Lesotho, the Maluti Mountains), has succeeded in adding this human dimension to the discussion through testimonies gathered from villages within the affected areas.

In the face of imminent resettlement – some to lowlands and semi-urban areas, some to other highland communities – people talk of their feelings of powerlessness and vulnerability, their distrust of the Lesotho Highlands Development Authority (the body responsible for implementing resettlement), and their fears for the future. Many express foreboding as to how losing their land will affect them in terms of not just livelihood but also self-esteem. Most of the men's experience of working in the South African mines has made them wary both of the dependence generated by being a wage labourer and of the finite nature of money. Mountain life might be frugal but, with land, they felt, they always had a productive resource – and a crucial degree of self-reliance. They speak with pride of their environmental knowledge and how it has enabled them to adapt and survive in a harsh landscape. But they also know that, when moving to urban areas, such skills are all but redundant. Source: Alton Byers, The Mountain Institute. abyers@mountain.org; Internet discussion. Primary sources used: The Lesotho Highlands Water Project webpage at: http://www.irn.org/programs/lesotho/index.asp?id=background.html; International Rivers Network Lesotho Campaign webpage at: http://www.lhwp.org.ls/projecthistory/project-description/project-description.htm; Panos Oral Testimony Program: Mountain Voices: Lesotho, the Maluti Mountains at: http://www.mountainvoices.org/lesotho.asp; Internet discussion.

In a number of developing countries, such as Bhutan and Laos, electricity generated in mountains is exported to earn foreign exchange. Compensation of the mountain communities for this resource use is non-existent or is grossly inadequate. This is illustrated by the San Gaban II scheme in the Cordillera Carabaya in Peru. This 110 MW facility, built by an international consortium of Peruvian, Brazilian, and French firms and largely funded by Japan, provides much-needed electricity to towns and industries in Southern Peru, including two multinational mines, which consume 30 per cent of the electricity. However, the scheme has yet to provide the electricity promised to most of the rural people in the Carabaya (Seimon 2001). This leads to the paradoxical situation of mountain settlements lacking electricity, but being bypassed by high-tension cables carrying electricity to main cities (E. Castro, Internet discussion). However, there are more encouraging examples, such as that of the Butwal Power Company in Nepal (a private enterprise), which now sells much of the energy generated by its 5.1 MW plant in Andhi Khola to the surrounding rural areas (B. Pandey, via MF-Asia Moderator, Internet discussion).

Small-scale hydropower development, on the other hand, is a promising approach to energy development in many mountain areas, especially in the developing world, with its large, untapped potential. In China, many mountain communities already rely on small-scale hydropower for most of their electricity. It minimizes social and environmental impacts, satisfies local needs, and is well accepted by local people. This can be illustrated by the example of the Mount Everest region in Nepal, where as many as 77 per cent of the local households used electricity after only one year of operation of a small local scheme, even though a power tariff was levied (Fischbacher 1999). Small-scale hydropower is cost effective, especially in difficult terrain with dispersed settlements, such as mountains, because costs of connection to local grids are low compared with those for large centralized plants, as no long transmission lines are required. In densely populated downstream areas, the costs of grid connection are roughly equal to the costs of electricity generation. In mountains with their difficult terrain and dispersed settlement, the costs of connection to regional or national grids can be two to five times as high as those of electricity generation in large, centralized plants (Rechsteiner 2001).

Experience has shown that small-scale hydropower benefits mountain populations in several ways. It powers local agro-processing mills, thereby reducing the drudgery of women by reducing the distance to the next mill. It replaces kerosene and other fuels with electricity for lighting: this has helped to extend working hours into evening or night, especially with regard to household, economic and social activities, and education. Small-scale hydropower now powers information and communication systems in remote areas, including radio, television, VHF for telecommunication services including education and training, and, in some places, computer services. In many places, it has encouraged the establishment, or growth, of micro-industrial enterprises that offer much-needed employment (Govinda Nepal, ITDG Nepal, Internet discussion).

The challenge in small-scale hydropower development is to find a viable balance between the size of the plant and the number and financial power of the users, so that the running costs (including servicing and smaller replacements) can be covered by the revenue levied from the users. (A.A. Sharma, contribution from email discussion). Other challenges include technical, managerial, and social aspects of development (see Case 4).

Case 4. Micro-hydropower at work: Experiences from Annapurna, Nepal

Because of steep gradients and abundant rainfall, the southern slopes of the Annapurna region in Nepal are especially suited for hydropower generation. In its efforts to provide energy alternatives to local communities, the Annapurna Conservation Project (ACAP) has installed 11 community-owned and -operated micro-hydropower plants in the area – including those at Sikles and Chhomrong, two villages in southern Annapurna.

The plant in Sikles, built in 1994, has been plagued with expensive landslide damage, high staff turnover, and the lack of savings. Electricity demand now exceeds supply. This has prompted the Village Electrification Committee (VEC), which operates the facility, to ban the use of low-wattage electric cookers and negate much of the firewood-saving effect of electrification. However, market penetration has reached 100 per cent and many positive social benefits (such as extended evening hours and reduced drudgery, particularly for women) have been noted. The plant in Chhomrong, built in 2000, has had fewer technical problems, but also suffers from an excess of demand over supply. As a result, although much firewood has been saved by its ban in tourist lodges, most households still cook with firewood, owing to the inadequate power supply, and income generation using electricity is limited mainly to tourist lodges. In addition, there is a communication gap between the Chhomrong VEC and the ACAP office: this is due, not least, to the fact that ACAP currently has no staff responsible for micro-hydropower, the single largest activity of the organization in financial terms. Despite these challenges, both plants show progress towards achieving ACAP's long-term goal of reducing the dependency on fuelwood and of development through community self-management. Source: E. Kim and B.S. Karky, 2002: Water resources use in the Annapurna Conservation Area. Internet discussion.

Fuelwood is by far the most important source of energy for a majority of mountain people worldwide, especially in developing countries. Reports on recent massive and widespread forest destruction to meet local fuelwood demand have often proved to be grossly exaggerated – for example, in the Himalaya and in the mountains of Eastern Africa, where reduction of forest cover is a much older process. Nevertheless, forests are under increasing pressure in many mountain areas, and women and children have to go increasingly further to collect wood. The situation is made worse by the low level of efficiency of fuelwood use – typically below 20 per cent. Moreover, the demand for space heating in mountains is greater than that in lowland areas: in the mountains of Nepal, for example, the energy required by households for heating is 56 per cent of household fuelwood demand, compared with 36 per cent in the lower hill areas (Rijal and Bhadra 2001). Governments and development agencies have recognized the problem of fuelwood and have tackled it from both the supply and demand sides. There are numerous afforestation schemes based on participatory approaches in mountain regions, and programmes to increase efficiency of fuelwood use have shown encouraging results where they are based on the needs of local users, especially women (see also Case 3). New affordable cooking devices and heating installations can double or triple fuelwood efficiency.

Traditional hydrocarbon energy sources, such as kerosene and gas, have their limitations in mountains, as they are not widely and reliably available owing to poor access and they are generally more expensive than fuelwood. Modern alternative energy sources, such as solar or wind power and biogas, have considerable potential in mountains as stand-alone facilities. Whereas the potential of solar power has been proven for many mountain areas, especially for high and dry regions such as the Tibetan Plateau and the southern-central Andes, the potential for wind power is still largely unknown. However, the technology – the fastest growing of all new energy alternatives – still encounters specific problems in mountain areas, such as wind turbulence and icing. Biogas also has its limitations, as it needs minimal temperatures for functioning. However, the greatest obstacle of all new energy alternatives is initial investment, which is still far beyond the means of poorer mountain communities. This has largely confined the spreading of these technologies to mountains in industrialized countries. Where they are used in mountains of the developing world, maintenance is a problem, owing to the lack of an effective and decentralized service network. Moreover, most modern energy alternatives do not produce large quantities of energy. In practice, this means that they have to be combined with other technologies, such as hydropower or fuelwood, to generate enough energy (S.S. Nyenhuys, Internet discussion, via MF-Asia Moderator).

Passive use of solar energy and insulation of buildings has probably not received the attention it deserves. It can significantly reduce energy needed for heating – especially in dry mountain areas and plateaus where solar energy is abundant, such as the Central Andes, parts of the Himalaya such as Tibet and Ladakh (see Case 5), Central Asia, and the Rocky Mountains. For example, recent studies in the mountains of Kyrgyzstan have shown that correct insulation with locally available material can reduce household energy demand by as much as 60 per cent. Combinations of alternative energy sources can help to increase the effectiveness of projects: many communities in the Rocky Mountains, for example, have recognized the importance of energy-efficient housing and require it of new buildings and renovations. Boulder, Colorado, USA is one of several communities that have adopted a "Green Building" section in their building codes, including options for renewable energy sources such as photovoltaic and solar thermal systems, as well as better insulation (L. Weddekind, Internet discussion).

Case 5. The use of passive solar energy for the construction of buildings and for income generation in Ladakh

Ladakh, in the Western Himalayan range of India, is a cold desert between 2,800 m and 4,500 m in altitude. Winters are very cold and temperatures can fall below −30°C. Trees do not grow under such conditions. During winter, local inhabitants use dung to cook and warm their houses. Roads are closed from mid-October to mid-May: the main town, Leh, can be supplied only by aircraft. On the other hand, the climate is very sunny. Simple solar technologies, based on passive solar use, local material, and locally trained experts, can help to create new economic activities (especially during the winter), such as greenhouses, poultry farming, and handicraft development – or, during the summer, such as solar dryers for fruit processing. For example, the investment required for a 50 m^3 greenhouse is US$450: this allows the production of more than 50 kg of vegetables per month.

The income generated is US$35 per month; the payback period is thus less than 2 years.

The potential for passive solar energy buildings in Ladakh is also high, for both public and private buildings. The investment is 10–20 per cent higher than that for a traditional building, but the reduction in fuelwood or dung consumption for space heating is 80–90 per cent. In peri-urban areas, the payback period of the extra building costs of a private building is less than 5 years. However, the diffusion of passive solar buildings is limited by cultural factors and by the challenge to retrofit double-storey buildings, the most popular type of house in Ladakh. Source: V. Stauffer, GERES, France (geres.stauffer@free.fr); Internet discussion.

Implications: Best practices

General principles

Adhering to the following principles will help to secure the sustainable development of access, communication, and energy in mountain regions:

- *Negotiate outcomes.* Adhere to the principle of negotiated outcomes involving all stakeholders, explicitly including local mountain communities. Negotiation must provide for independent arbitration, where necessary, and give equal weight to environmental, social, and economic aspects of development.
- *Share benefits through the principle of equity.* Compensate mountain regions for services provided to society by sharing benefits that accrue nationally or globally. There is less need for additional external funding than for more equitable re-allocation of existing funds within countries.
- *Tailor development to mountain regions.* Respect the specificities of mountain communities and environments by applying or developing appropriate and non-stereotypical technical solutions in the transport and energy sectors. The keywords are decentralization, protection from natural hazards, and careful construction to avoid damage to fragile environments. Phased development of facilities can be useful as it offers the possibility to make adjustments in order to mitigate undesired impacts (A. Sharma, A. Thomson, Internet discussion).
- *Build on existing facilities and experiences.* Optimize use of existing facilities and institutions by improving their efficiency and effectiveness before creating new ones. Promote and enforce demand-side management in energy use, especially in well-developed mountain regions.

Best practices: Energy

With regard to energy production in mountains, best practices can be characterized by approaches focusing on decentralized and small-scale development. This is especially true for hydropower generation, for which there is still a large untapped potential in many mountain regions in developing countries. With regard to small-scale hydropower, plant ownership and sound technical solutions are important points to be considered: second-best equipment is often not enough and hence not appropriate – even for micro schemes (Mor and Zimmermann 2001). Larger facilities may also be necessary to supply growing urban and industrial centres in mountains and downstream areas and to reduce current high consumption levels of non-renewable energy such as fossil fuels. Where such facilities are built in mountain areas, they must be based on impact assessments that show that their effects are not disruptive to the local society or harmful to the environment. It should be mandatory for local mountain communities to be supplied with the electricity generated at preferential rates. Facility operators should pay a water royalty, of which a specific share is returned to the local mountain communities as compensation for the use of mountain waters and for the protection of watersheds. This is possible also in non-industrialized countries, as illustrated by the example of Costa Rica, where private landowners are compensated for environmental services (see ch. 4, this volume). The funds largely come from a tax levied on fossil-fuel consumption and thus there is no additional burden to the national treasury (Campos and Calvo 2000). The shift from supply-side energy policies to a demand-oriented approach to electricity generation is of fundamental global importance.

With regard to fuelwood use, best practices include reducing fuelwood consumption by increasing the efficiency of fuelwood use in close cooperation with user groups and in consideration of the greater need for space heating in mountain areas than in the lowlands. In the same vein, passive solar energy use – and, especially, insulation of buildings – have a large untapped potential, the harnessing of which has to be included in best practices.

Modern alternative energy technologies such as wind and solar power are included in best-practice approaches, as running costs are low and the resources (wind and solar energy) are unlikely ever to run out. As initial investment costs will remain high in future, such installations should be mainly considered for public infrastructure such as schools and health facilities, where government funds and external funds from the donor community are available. Best practices include setting up mechanisms for servicing and maintenance of these sophisticated facilities.

Best practices: Access and communication

With regard to access, roads will remain the most important means for improving access to mountain areas in many parts of the world. Best practices include labour-intensive approaches in developing countries. In industrialized countries, where this approach is not feasible owing to high labour costs, excavators rather than bulldozers should be used wherever possible in mountain areas, as these cause less damage to the environment. Best practices in transport development in mountain areas also include a cost–benefit analysis of various alternative forms of access, such as roads or ropeways. When it comes to priority setting in road construction, best practices use a list of criteria including poverty level, number of people benefiting, per capita cost of construction, remoteness, and environmental vulnerability, along the lines of the criteria developed, for example, by the Social Fund for Development in Yemen, or by the Green Road Concept (GRC) developed in Nepal (see Case 6).

Case 6. Green Road Concept, Nepal

The Green Road Concept (GRC) emerged in the Nepal Himalayas from lessons learnt from mountain-road projects initiated in the early 1970s. The concept was elaborated by analysing about a dozen mountain-road projects. A list of best practices was compiled and summarized under the label of GRC. The label is based on a series of principles including cultural identity of mountain populations, community-based public work, strong geological dynamics, and fragile ecosystems including great diversity of rainfall, as well as limited financial resources at local and national levels. It also includes the option of staged upgrading with increasing traffic, especially concerning road surface.

Specifically, the principles include the following: participatory planning; adoption of environmentally friendly road technologies; promotion of labour-based technologies in order to generate local income, particularly for the poorer sections of the population; application of performance-based work management, considering both quality and quantity measurements; specification of road ownership based on subsidiarity considerations; adaptation of the work plan for road construction to agricultural off-periods; encouragement of gender equity in decision-making and work provision; promotion of local capacity building to facilitate future maintenance works; and combination of funds from local and central governments, which are complemented by donor support if necessary. Regular public audits are important to ensure financial transparency, optimize cost efficiency, and increase the confidence of the local population with regard to utilization of funds. An agreed-upon road-maintenance scheme, including local institutions supporting it, is also worked out during the construction phase. Source: Werner P. Meyer, Rural Development Programme GTZ, PGO Box 1457 Kathmandu; Internet discussion.

Best practices take care to minimize direct environmental impacts of the construction of transport infrastructure, to prevent erosion, to restore vegetation damaged during construction, and to secure slope stability. Engineering biology can help reach these aims. All these factors increase the costs of constructing and maintaining roads, railways, and other means of transport in mountain areas, for which best practices should make ample provision – a continued challenge for new projects at local, national, and international levels (see Case 7).

Case 7. Transport corridors in Central Asia: Development and conservation in a transboundary context

The mountains of the Pamir, Tien Shan, and Altai regions in Central Asia are increasingly being drawn into globalized transport networks, which changes their status from regional culs-de-sac to gateways. The main reasons for this change are the disintegration of the Soviet Union and the creation of new independent states, and China's plans to promote the development of its western provinces. According to current plans, at least five transboundary mountain regions will be affected by road and rail construction, which include such far-reaching projects as the Tranceco railroad route from the east coast of China to Rotterdam, and the railway cum gas-pipeline link between Xinjiang and Siberia.

The challenge for these projects is to reconcile much-needed local development with nature conservation. The mountains of Central Asia harbour great biological and ecosystem diversity, and include a UNESCO–World Heritage Site in the Altai Mountains. Biodiversity has been eroded through such illicit means as poaching, illegal hunting, and overgrazing – a sad reflection of the region's grave economic problems. There is thus a need to formulate strategies and action plans for sustainable development based on transboundary cooperation. Source: Yuri P. Badenkov, Institute of Geography, Russian Academy of Science, Moscow.

Modern communications technology will become increasingly important in future and should, therefore, be promoted. However, initial costs are still prohibitive to many people living in mountain areas, especially in the developing world. Best practices therefore start with providing the requisite equipment and human capacity to decentralized institutions in mountain areas – such as schools, health facilities, local government and NGOs, or local post offices – which generally have less difficulty than private households in securing the means for maintaining the requisite infrastructure and which can provide access to the facilities to surrounding populations. The basic idea is to establish decentralized service centres that offer telephone, email, and internet facilities against an affordable service fee. Financial support, such as finance schemes, could greatly facilitate the establishment of such centres, especially if initiated by NGOs, local groups, or private-sector initiatives (P. Helmersen, Internet discussion). Best practices also include an element of awareness creation about the opportunities of modern information technologies, including training of interested user groups and including (English) language skills, where necessary.

Best practices in improving access in mountain areas consider the potential of animal transport, especially as a complement to motorized transport (village-to-village and village-to-road transport). Animal transport provides a cost-effective form of moving goods and people in many mountain regions, is affordable to many, and is integrated in local livelihoods.

Linkages

Energy

Important linkages include:
- national and regional energy development policies;
- urban development, development of smaller towns in mountains;
- promotion of small-scale industry in mountain areas;
- promotion of employment in secondary and tertiary sectors in general in mountain areas;
- building standards (energy saving) and clean energy promotion;
- long-term and short-term credit;
- development of communications and access infrastructure in mountains;
- good governance, decentralization, and compensation mechanisms for use of mountain resources;
- equity and ownership of resources;
- sustainable use of forest resources in mountains;
- watershed management and protection of headwater areas;
- initiatives keyed to tackle man-made causes of global warming.

Access and communications

Important linkages include:
- mountain-specific regional development policies;
- infrastructure development (health, education), and physical planning, especially in larger villages and towns in mountains;
- employment generation;
- migration;
- promotion of local and regional markets, and marketing of mountain products;
- tourism development and amenity migration;
- forest management and protection of forest resources;
- industry, mining, and hydropower generation;
- transboundary cooperation in mountain development.

Key actions

Access, communications, and energy are powerful agents of change; the changes that they induce and spearhead must be managed carefully. Access, communications, and energy are key topics to be considered in mountain conventions, charters, and transboundary agreements at regional levels.

Downstream interests have largely dominated energy and transport development in mountain areas. In a rapidly urbanizing and increasingly globalized world, mountains continue to play an important role in securing transport links and providing energy for surrounding lowlands and urbanized areas. Providing such services often has negative impacts on mountain regions. It is critical that mountain communities benefit from both the diverse energy resources available in mountain environments and the development of transport networks and other new means of communication. A careful balance between downstream interests and mountain interests is urgently required. In general, the development of access, communications, and energy production requires considerable technical and managerial expertise and involves substantial costs; capacity building and training among producers and users of such facilities is thus crucial. Likewise, financing is a key issue in policies and programmes in the energy, communications, and transport sectors.

Key actions for local communities

Mountain populations should be ready to participate in the development of energy, access, and communication facilities that serve local needs. This may require the (re-)establishment of local institutions to shape local opinion and secure ownership of installations. Taking advantage of improved access and communications technologies, mountain communities can greatly benefit from establishing links between themselves and creating regional institutions to enhance their political position and to promote tourism or the marketing of local products. Local or regional communities can also substantially improve their standing through linkages with institutions in civil society at large, such as international and national NGOs and the media.

Key actions for national governments and authorities

National authorities must formulate sectoral policies in energy, transport, and communications that acknowledge the need for mountain-specific approaches and technologies. Modernization should be promoted where it not only benefits outside interests but also supports mountain people, and should take account of traditional forms of transportation and energy use, so as to transform rather than disrupt mountain livelihoods. National authorities should enforce safety standards for dams, roads, railways, and ropeways in mountains; quality standards for installations and devices for communications and energy generation and use; environmental standards for minimum water flow and protection against natural hazards; and political and social standards such as full compensation when negotiated outcomes infringe on, or cause loss of, property rights for reasons of regional or national interest. Governments should introduce time-bound licences for hydropower facilities and water royalties to be paid by utilities and should return a negotiated share of the revenue generated to compensate mountain communities for the goods and services they render to society at large. The global concern over climate change could provide a new source of funding for clean energy production in mountains, such as hydropower, solar, or wind power.

Governments could use revenues from taxes on fossil fuels or from mountain tourism (e.g. trekking permits, entrance fees) to promote the use of renewable energy and to help mitigate negative impacts caused by access and communication development in mountain areas. Much more than has been the case in the past, governments should encourage the private sector to participate in developing appropriate technologies and tools for the energy and access/communications sectors in mountain areas. Specific policies to achieve this aim could include tax reduction and, especially, the establishment of credit schemes for the implementation of facilities of access, communication, and energy generation, and for technology development; typically, commercial banks are seldom interested in providing credit for infrastructure development in mountains, owing to low return on investment and long payback periods.

Key actions for civil society and NGOs

Civil society and NGOs must be involved in any negotiations to establish a new project. There is a need (a) to see that authorities, facility operators, and private enterprises comply with approved policies, agreements, and standards; (b) to expose violations; and (c) to lobby decision makers, politicians, or the mass media to initiate remedial action. Civil society and NGOs can support appropriate projects and approaches in energy and communications development in mountains, and can create awareness of best practices for infrastructure development by disseminating information about successful initiatives. NGOs are also able to carry out holistic projects, which integrate (for example) energy production with income generation and private-enterprise development at the local level, while at the same time involving local governments to achieve larger-scale coverage (B. Pandey, via MF-Asia Moderator, Internet discussion). They have a crucial role to play in supporting mountain community networks and unions, enhancing the position of mountain people in negotiation processes, and helping to prevent undesired outcomes and projects.

Key actions for international organizations and the donor community

These institutions should provide financial support solely to projects based on the principles of sustainable development, as a result of negotiated outcomes and agreed processes involving all relevant stakeholder groups, including local interests. Guidelines need to be established for supporting local and regional community networks and for agreements that promote regionally balanced development, including social and environmental standards, in the energy and transport/communications sectors. International organizations and the donor community should support transboundary cooperation for the sustainable development of international mountain areas, e.g. with regard to transit corridors, watershed management, hydropower generation, and electrification.

Key actions for the private sector and professional associations

Private-sector enterprises have a responsibility to support development that includes environmental and social as well as economic considerations. Together with professional associations, these enterprises should develop voluntary codes of conduct for energy and transport development in mountain areas. They should (a) comply with regulations and standards established by international and national authorities; (b) train their staff in, and make executives aware of, positive and negative impacts of transport and energy development in mountain areas; and (c) abide by the provisions of anti-bribery conventions. At national or regional levels, they should set up business councils that encourage and promote the development of technologies in the communications and energy sectors that are adequate for mountain areas. Moreover, the private sector has a responsibility to help to introduce – and maintain – quality standards. This includes, especially, appropriate technology – which has adopted the right intention and approach but has produced too many poorly functioning, non-durable products and installations. Appropriate technology does not, therefore, necessarily involve using low-cost materials and manufacturing technology (S. Nienhuys, Internet discussion, via MF-Asia Moderator).

Key actions for the scientific and research communities

As demand for energy increases in all regions, including mountain areas, there is an urgent need for innovation leading to more efficient use of existing technologies and to new technologies in energy and transport. Researchers should join forces with private enterprises, professional associations, and mountain representatives to develop realistic and acceptable solutions and to test new devices and approaches. However, innovation is not limited to technology: the scientific and research communities can help to formulate compensation mechanisms (e.g. water royalties) and quality labels (e.g. green electricity), or can help to define enabling incentives for new approaches and technologies. They can support policy makers in the formulation of a resource policy that gives priority to renewable resources.

The impacts of energy, communications, and access development often remain poorly understood. Short-term impacts may differ substantially from long-term impacts. Monitoring of impacts – including technical, socio-economic, and environmental aspects – is essential. Monitoring and documentation provide valuable information for informed decision-making relating to new projects and also provide data for designing measures to mitigate the unresolved legacies of past projects.

Going beyond 2002

The International Year of Mountains 2002 presents an excellent opportunity for further development of long-term collaboration between all the stakeholders concerned with access, communications, and energy in mountains, to ensure that these key sectors benefit both mountain people and those dependent on them in surrounding lowland areas.

Note on references

This chapter is based largely on material drawn from the following publication: *Mountains of the world: Mountains, energy, and transport*, edited by the Mountain Agenda, and published in 2001 for the UN-CSD. The publication was commissioned, and largely funded, by the SDC (Swiss Agency for Development and Cooperation). Specifically, this chapter has drawn on contributions by the following authors in the above (2001) publication: Bikash Pandey, Rudolf Rechsteiner, Kamal Rijal and B. Bhadra, Martin Price and C.S. Houston, Anton Seimon, Udo Schickhoff, Urs and Ruth Schaffner, Peter Hartmann, Franz Gaehwiler, Hans Dieter Schmoll, and Paul Starkey.

All references to Internet discussions refer to material provided via the Mountain Forum e-consultation (see http://www.mtnforum.org/bgms/ schedule.htm).

BIBLIOGRAPHY: REFERENCES

Badenkov, Y.P. 2002. "Altai Mountain Knot – Development versus conservation? Syndromes of globalisation." Case study on Mountain infrastructure: Access, communications, energy. Mountain Forum e-consultation for the UNEP/Bishkek Global Mountain Summit.

Byers, A.C. 2002. "The Lesotho Highlands Water Project: Supporters, critics, and mountain voices." The Mountain Institute. Case study on Mountain infrastructure: Access, communications, energy. Mountain Forum e-consultation for the UNEP/Bishkek Global Mountain Summit.

Campos, J.J., and J.C. Calvo. 2000. "The mountains of Costa Rica: Compensation for environmental services from mountain forests." In Mountain Agenda, *Mountains of the world: Mountain forests and sustainable development*. Berne: Centre for Development and Environment.

Fischbacher, C. 1999. *Entwicklung durch Technik*. Vienna: OEFSE, Oesterreichische Forschungsstiftung für Entwicklungshilfe.

Gaehwiler, F., and M.N. Lamichaney. 2001. "Bridges for rural access in the Himalaya." In: Mountain Agenda, *Mountains of the world: Mountains, energy, and transport*. Berne: Centre for Development and Environment.

Hartmann, P. 2001. "Human power instead of machines: Rural access roads in West Flores, Indonesia." In: Mountain Agenda, *Mountains of the world: Mountains, energy, and transport*. Berne: Centre for Development and Environment.

Kim, E., and B.S. Karky. 2002. "Water resources use in the Annapurna Conservation Area: Case study of micro-hydropower management in Sikles and Chhomrong." Case study on Mountain infrastructure: Access, communications, energy. Mountain Forum e-consultation for the UNEP/Bishkek Global Mountain Summit.

Meyer, W.P. 2002. "Green Road Concept for Mountain Road System." Paper contribution on Mountain infrastructure: Access, communications, energy. Mountain Forum e-consultation for the UNEP/Bishkek Global Mountain Summit.

MF-Asia Moderator. 2000. Case study adapted from H. Kreutzmann: "Improving accessibility for mountain development: Role of transport networks and urban settlements." In: M. Banskota, T.S. Papola, and J. Richter (eds). *Growth, Poverty Alleviation, and Sustainable Resource Management in the Mountain Areas of South Asia*. Proceedings of the International Conference, 31 January–4 February 2000, pp 492–499.

Mor, C., and W. Zimmermann. 2001. "Developing small-scale hydropower in Nepal." In: Mountain Agenda, *Mountains of the world: Mountains, energy, and transport*. Berne: Centre for Development and Environment.

Pandey, B. 2001. "Mountains and energy: Mountains as global centres of hydro-power." In: Mountain Agenda, *Mountains of the world: Mountains, energy, and transport*. Berne: Centre for Development and Environment.

Price, M.P., and C.S. Houston. 2001. "Human energy in the mountains." In: Mountain Agenda, *Mountains of the world: Mountains, energy, and transport*. Berne: Centre for Development and Environment.

Rechsteiner, R. 2001. In: Mountain Agenda, *Mountains of the world: Mountains, energy and transport*. Berne: Centre for Development and Environment.

Rijal, K., and B. Bhadra. 2001. "Sustainable fuelwood use in mountain areas." In: Mountain Agenda, *Mountains of the world: Mountains, energy, and transport*. Berne: Centre for Development and Environment.

Schaffner, U., and R. Schaffner. 2001. "Access road construction in Ethiopia and Yemen." In: Mountain Agenda, *Mountains of the world: Mountains, energy, and transport*. Berne: Centre for Development and Environment.

Schickhoff, U. 2001. "The Karakorum Highway: Accelerating social and environmental change in a formerly secluded high mountain region." In: Mountain Agenda, *Mountains of the world: Mountains, energy, and transport*. Berne: Centre for Development and Environment.

Schmoll, H.D., and D. Seddon. 2001. "Ropeways for mountain tourism and development." In: Mountain Agenda, *Mountains of the world: Mountains, energy, and transport*. Berne: Centre for Development and Environment.

Seimon, A. 2001. "Mountains and transport." In: Mountain Agenda, *Mountains of the world: Mountains, energy, and transport*. Berne: Centre for Development and Environment.

Starkey, P. 2001. "Animal power: appropriate transport in mountain areas." In: Mountain Agenda, *Mountains of the world: Mountains, energy, and transport*. Berne: Centre for Development and Environment.

Stauffer, V. 2002. "Income generation activities using solar energy in Ladakh Western Himalaya Range, India." Case study on Mountain infrastructure: access, communications, energy. Mountain Forum e-consultation for the UNEP/Bishkek Global Mountain Summit.

Widdekind, L. 2002. Encouraging sustainable energy in Colorado's Rockies. Boulder Energy Conservation Center, Colorado. Case study on Mountain infrastructure: Access, communications, energy. Mountain Forum e-consultation for the UNEP/Bishkek Global Mountain Summit.

World Commission on Dams. 2000. "Dams and development: A new framework for decision-making." Report of the World Commission on Dams. An Overview – 16 November 2000. https://www.dams.org/report/wcd_overview.htm.

BIBLIOGRAPHY: INTERNET DISCUSSIONS

Contributors are listed below in order of their citation in the main text. NB: Authors of Internet case studies are listed in the reference list above.

Montgomery, Layton L.; title of contribution: no specific title; affiliation: PhD candidate, Centre for Research Policy, University of Wollongong, Wollongong, Australia; date of discussion: 10 March 2002.

Niewiadomski, Zbigniew; title: "Hutsul Horse Reintroduction in Bieszczady National Park, Poland"; affiliation: none given; date: 7 March 2002.

Pandey, Bikash; title: no specific title; affiliation: none given; date: 14 March 2002.

Nepal, Govinda; title: "Technology for Mountain People: A Glimpse of ITDG Nepal's Efforts"; affiliation: Advisor to ITDG Nepal (i.e. Nepal Office of Intermediate Technology Development Group Nepal Office); date: 1 April 2002.

Sharma, Ajay; title: no specific title; affiliation: Doctoral Fellow PSG, Central Queensland University, Rockhampton, Australia; date: 6 March 2002.

Njenhuys, Sjoerd; title: none; affiliation: Consultant EPA; date: 21 November 2001.

Thomson, Alan; title: none; affiliation: Senior Research Scientist, Canadian Forestry Service, Pacific Forestry Centre, Victoria, Canada; date: 6 March 2002.

Helmersen, Per; title: none; affiliation: Senior Research Psychologist, Telenor R&D, Norway; date: 14 March 2002.

4

Legal, economic, and compensation mechanisms in support of sustainable mountain development

Maritta R.v. Bieberstein Koch-Weser and Walter Kahlenborn

Summary

Environmental service agreements are urgently needed, in the face of observable, global trends towards environmental degradation in mountain areas. Region-specific approaches need to be developed for the valuation and contracting of upstream environmental services by downstream communities and enterprises who depend on reliable quantities of good-quality water, and for disaster prevention.

This chapter recommends the development of region-specific mechanisms and agreements. As a point of departure for the eventual development of specific instruments and regional agreements, it provides an overview of prominent current examples and cases on which the development of tools for the valuation, negotiation, implementation, and monitoring of environmental services could build.

In its last section, the chapter provides operationally oriented guidance for the planning of systems and agreements for downstream–upstream payments for environmental services.

Issues

The need for environmental service agreements

Sustainable water development and the mitigation of natural disasters in entire river basins depend in large measure on the ways in which up-

63

stream water sources and soils in mountain areas are protected. Environmental services provided by mountain regions are often noticed only when they are lost, as in the case of downstream floods caused by upstream deforestation. As half of humanity depends on fresh water that originates in mountain watersheds, solving these problems is critical for global environmental security (Mountain Forum 2001). However, in most regions of the world, downstream people have no tradition of negotiating environmental safeguards with mountain folk upstream, nor do they have legal and economic instruments and social-organization models to do so.

To date, when making land-use decisions, upstream dwellers have generally not taken the value of environmental services provided by their forests and other permanent soil-protection vegetation into account, because they normally do not receive any compensation for these services. Nor will they invest in soil-conservation practices to protect watersheds to the benefit of downstream neighbours.

As a result of the lack of attention to mountain watersheds, there is a dangerous trend of accelerating erosion in catchments at the source and dwindling water availabilities downstream. Around the world, one can observe a lack of effective, long-term, downstream–upstream environmental maintenance and compensation agreements. Ironically, while the global population has tripled over the last century and water scarcity is already acute in many parts of the world, the protection of upstream water sources has become worse, not better.

Environmental service agreements are now urgently needed, in the face of observable, global trends towards environmental degradation in mountain areas. Region-specific approaches need to be developed for the valuation and contracting of upstream environmental services by downstream communities and enterprises dependent on reliable quantities of good-quality water, and for disaster prevention.

Downstream effects of environmental mismanagement

The impact of the degradation of mountain ecosystems through clear-cutting and unsustainable forestry and agricultural practices can be tremendous and costly downstream (Hamilton and Bruijnzeel 1997; Hamilton, Gilmour, and Cassells 1997; Jodha 1997). Impacts include shallower aquifers and wells, siltation of hydropower and irrigation reservoirs through hillside erosion, less water retention in the dry season, and more violent floods in the rainy season. Water quality suffers from agricultural runoff, which spoils the purity of renewable sources of fresh water, or – with changes in the overall water level – from increased levels of salinity, arsenic, and other substances. Loss of mountain forest cover and the subsequent erosion account for increases in natural hazards, such as avalanches, landslides, and floods. Floods and mudslides that start in deforested mountain ranges cause by far the most costly damage: globally, the total damage to property and infrastructure accounts for tens of billions of dollars every year (International Year of Mountains 2002).

The plight of mountain dwellers

The protection of water sources and mountain watersheds depends on people. Mountain communities tend to be comparatively poor and isolated. In many destitute mountain regions, the inhabitants' lives leave no room for choosing the environmental high ground; instead, they will work any land – no matter how fragile – in their struggle for sheer short-term survival. In many instances, traditional practices – which may have guaranteed sustainable use in past centuries – have made way for unsustainable land-use patterns. Populations have outgrown the carrying capacity of the land, and have moved on to increasingly more fragile, steep lands for farming and livestock husbandry. Also, for instance in the Andes, former lowland populations – entirely inexperienced in mountain farming – are now being pushed into mountain regions in their quest for subsistence agriculture.

The fragility of mountain environments

The incidence of environmental degradation in mountain areas is especially high, because of their extreme fragility. Mountain ecosystems are characterized by steep slopes resulting in both rapid and gradual geomorphological processes, and low temperatures, which cause vegetation growth and soil formation to occur very slowly. Soils are usually thin, young, and highly erodible. Under these conditions, farming in marginal mountain areas easily causes environmental imbalance. Once eroded, mountain areas may need hundreds of years to recover (Byers 1995).

The world's remaining mountain forests are essential for minimizing soil erosion. They still cover more than 9 million km^2 with almost 4 million km^2 above 1,000 m. They represent 28 per cent of the world's closed forest area (Kapos et al. 2000). People benefit from mountain forests in many ways: in general, forests slow the rate of runoff in a watershed, ensuring a certain base flow and minimizing flooding in small watersheds; they also reduce soil erosion and they can improve water quality (Johnson, White, and Perrot-Maître 2001). Yet, despite the benefits that mountain forests provide in terms of the overall environmental regime, they have been disappearing at a startling, unprecedented rate in the last decade. Reasons for deforestation are manifold: it tends to be driven by population growth, uncertain land tenure, inequitable land distribution, illegal logging, and the absence of strong and stable institutions (Hamilton, Gilmour, and Cassells 1997).

Combined, these social and institutional factors cause increases in settlements, agriculture, and livestock developments in unsuitable, fragile, mountain areas. Their effects can be exacerbated by a simultaneous excessive development of infrastructure, such as that for tourism and recreation, or for logging. As soil erosion increases, it turns into a driver by itself, propelling farmers into patterns of shifting cultivation or as migrants on to yet further new settlements.

Payments for environmental services

One promising instrument for downstream–upstream cooperation is payments for environmental services (PES). Water users compensate the watershed's upstream forest owners and landholders for, for example, forest conservation or reforestation or other services to maintain or improve water quantity and quality downstream. By giving an economic value to the environmental services provided by, for instance, the maintenance of forests, ecosystem protection can become an attractive alternative to other land uses pursued by the forest owners. To date, there are only a few cases worldwide that involve PES: most are in South America, but they are not unknown in other regions of the world. For this chapter, those cases were considered that involve compensation schemes between downstream beneficiaries and upstream suppliers of environmental services in mountain regions.

The cases briefly described in the next section of this chapter seem to follow some of the same principles. PES require as elements especially the following:

- valuation of the environmental services, from the vantage point of one or several stakeholder groups downstream;
- social organization effective enough among the respective upstream and downstream negotiating parties to allow for tangible payment agreements;
- clear and verifiable agreement on targets, and related implementation and monitoring arrangements;
- a legal and institutional framework;
- provisions for conflict resolution.

The regional dimension

Most of the world's major rivers traverse several provinces and/or countries. Transboundary environmental-management agreements and PES will therefore remain pivotal for the future. Transboundary river-basin management schemes exist, for instance for the Mekong and Danube rivers; however, because of their complexity, these have not been included among the case material in this paper.

The global dimension?

In addition to the examples of PES presented in this chapter, there may be additional opportunities for mountain forests under global carbon offset schemes under the Clean Development Mechanism (CDM) associated with the Kyoto Protocol on Climate Change. In addition to climate and carbon sequestration, CDM applications – especially in the poorest mountain regions of the world – could bring great benefits in terms of watershed protection, poverty alleviation, and disaster mitigation.

Case studies: Payments for environmental services

Case 1. Australia: Irrigators finance upstream reforestation

Background

One of the main problems in Australia is the salinization of land as a result of deforestation. In this particular environment, the water contains increasing amounts of dissolved mineral salts. In the Murray–Darling watershed, the Government Agency entitled State Forests of New South Wales (SF) is responsible for sustainably managing the forests. The Macquarie River sub-watershed, which is particularly vulnerable to salinity owing to its physical characteristics, is also particularly affected by salinization due to land clearing (Coram 1998; Perrot-Maître and Davis 2001).

Participants in the scheme

Participants in the scheme are SF and Macquarie River Food and Fibre (MRFF), an association of 600 irrigation farmers in the Macquarie River catchment. MRFF pays for the environmental service that is provided by SF and private upstream landowners, who represent the third party of the scheme.

The scheme itself: Legal and economic procedures

In 1999, SF entered into a Pilot Salinity Control Trade Agreement with MRFF, according to which MRFF pays the agency to replant trees in the upper catchment area. This public–private partnership works as follows. The irrigators pay about US$42 per hectare of reforested land per year for 10 years to SF, purchasing transpiration or salinity-reduction credits which were earned before by the agency through reforestation of 100 hectares of land. SF uses the revenues of this trading scheme to replant more trees on public and private lands. Private landowners receive an annuity, but the forestry rights remain with SF. The ambitious aim is to restore 40 per cent of the cleared forest, which is necessary to reverse the salinity process (Perrot-Maître and Davis 2001).

The agreement does not yet represent a real trading scheme because only two partners take part in the trade. So far, there have been few problems with implementation. Because it was mainly intended as a trial of the use of a market-based approach to help control dry-land salinity, it has already provided valuable insights into the working of such a scheme (e.g. possible buyers and sellers, definition of the product). If a full trading scheme is ever to be implemented, one has to deal with the fact that causes and effects are difficult to determine and that it is, therefore, not easy to predict the improvement in water quality downstream that will result from a lower water table due to increased transpiration in the upstream area. One also has to deal with the "free-rider" phenomenon, i.e. that MRFF is paying to achieve the benefits of improved water quality downstream, whereas all other water users receive those benefits for nothing (Salvin, S. 2002, pers. comm.).

Case 2. Colombia: Irrigators pay upstream landowners for improvement of stream flow

Background

In the extremely fertile Cauca River basin of Colombia, water became scarce in the summer whereas floods were experienced in the rainy season. Furthermore, in the late 1980s, rapid urban, industrial, and agricultural development resulted in sedimentation in the irrigation channels. Farmers were especially affected by the problems because Colombian laws require that domestic users are provided with water first.

Participants in the scheme

In order to tackle these problems, in the late 1980s and the early 1990s farmers formed more than 12 water-user associations in the different sub-watersheds and decided to pay upstream forest landowners for the management of their forests. The third participant in the scheme was the public Cauca Valley Corporation (CVC), the regional environmental authority that has been responsible for water allocation and the protection of the resources within the area since 1959. The CVC manages the fund.

The scheme itself: Legal and economic procedures

The farmers make voluntary payments to the CVC, which places contracts with upstream forest landowners dealing with reforestation, erosion control, and spring and stream protection according to sub-watershed-management plans. Furthermore, the public–private fund is used for land acquisition and economic development in upland communities. Because, in Colombia, private associations are not legally authorized to implement watershed-management plans, this is the only possible setting.

The association members voluntarily pay an additional water-use fee of

US$1.5–2/litre on top of an already existing water-access fee of US$0.5/litre (Perrot-Maître and Davis 2001; Tognetti 2001). Between 1995 and 2000 (with the year 2000 considered a low point because of the economic crisis in Colombia), a total investment of over US$1.5 billion represents a rough, conservative estimate. Unfortunately, information concerning the amounts of the funds since the associations were formed has not been systematically collected.

To date, there have been no problems regarding the implementation of the scheme: communities were highly motivated to take part in watershed-protection measures. Concerning the effects of those measures, no study has been undertaken to determine if, for example, increased upstream land cover has had an effect on water flow. Although there was less flooding between 1988 and 1998, this could have been due to milder weather conditions. In two sub-watersheds in the region, increased water flow has been seen during the dry season (Echavarría 2002).

Case 3. Costa Rica: Hydroelectric companies pay upstream landowners via FONAFIFO

Background

From 1950 to 1983, Costa Rica's forests were reduced to 49 per cent of their previous area because of clear-cutting in order to plant coffee, bananas, and sugar. The forests left were mostly in protected areas. In 1996, the new forestry law was approved, aiming at encouraging conservation through PES provided by forests.

The PES Program was intended to maintain forest cover through the provision of compensation to forest owners for the benefits they produce. To operationalize the PES Program, in 1997 the Government of Costa Rica established the National Forest Office and National Fund for Forest Financing (FONAFIFO) within the Ministry of the Environment, which is primarily financed through a 5 per cent sales tax on fossil fuel. FONAFIFO pays forest owners for 5 years for the mitigation of greenhouse-gas emissions and the protection of watersheds, biodiversity, and scenic beauty (Rojas and Aylward 2002).

Landowners who protect their forests receive US$45/ha/year; those who sustainably manage their forests receive US$70/ha/year; and those who reforest their land receive US$116/ha/year. In the second and third cases, plans must be generated by professional foresters (Rainforest Alliance 2001). Although most deals are made between FONAFIFO and upstream forest owners, private companies, especially in the hydroelectricity sector, have also initiated contracts and have become partners in PES schemes.

Participants in the scheme

Besides the public FONAFIFO (which, in these cases, serves as a mediator between the contracting parties) there are two other partners in the voluntary PES Program – public or private hydroelectric companies, who pay for the service, and upstream forest owners, who provide it. By the end of 2001, agreements had been negotiated with Energía Global de Costa Rica (Sarapiqui Watershed), Hydroelectrica Platanar (located in San Carlos), and the Compania de Fuerza y Luz, which distributes electricity in the capital San José, as the downstream partners (Perrot-Maître and Davis 2001). In the cases of Energía Global and Hydro-electrica Platanar, the NGO FUNDECOR served as a facilitator. It mainly supported forest owners who wanted to be included in the PES Program, by negotiating with FONAFIFO (FUNDECOR 2001).

The scheme itself: Legal and economic procedures

FONAFIFO serves as a mediator between the contracting parties and provides an institutional, standardized framework for compensation payments. The hydroelectric companies make their payments to the Fund, which negotiates with and pays the upstream landholders, often represented by FUNDECOR. The reason why hydroelectric companies are interested in upstream forest conservation and thus take part in the Program is that the protection of water resources is important for the effective and efficient working of the hydroelectric plants (increased regularity of stream flow and reduction of reservoir sedimentation) (Perrot-Maître and Davis 2001).

The first hydroelectric company seeking to protect its watershed was Energía Global de Costa Rica, which operates two hydroelectric dams (Don Pedro, Río Volcán). This private company pays 40 upstream landowners for reforesting their land, adopting sustainable forestry techniques, and/or the preservation of their woods. Energía Global pays US$18/ha/year to FONAFIFO – which adds another US$30/ha and then makes cash payments to land owners. FUNDECOR controls the implementation of the conservation activities and manages the legal and administrative operation. The sum of US$48/ha/year mainly equals potential revenues from cattle ranching (FUNDECOR 2001).

The hydroelectric company Hydroelectrica Platanar pays US$30/ha/year to FONAFIFO, which adds a certain amount and pays upstream forest owners for the voluntary inscription of their properties in a forest regime, which includes the implementation of a Management Plan that guarantees the unchanged survival of the forests (FUNDECOR 2001; Rojas and Aylward 2002).

The National Power and Light Company (Compania de Fuerza y Luz) pays US$45/ha/year to FONAFIFO for forest management, conservation, or reforestation projects as well as the promotion and follow-up of such projects in its watershed (Perrot-Maître and Davis 2001).

Although, overall, the PES Program can be seen as a success, there is some criticism of the fact that only a few women and indigenous communities are enrolled. The reason for this is that enrolling in the PES Program is expensive because professional foresters must be hired to gather the required information. Here FUNDECOR comes in: the NGO seeks contact with upstream forest owners and carries out the technical studies of their properties; other NGOs handle the studies and paperwork for a fee. Furthermore, the fact that companies who manage their forests receive more money than those who truly protect it, has been criticized. Problems regarding implementation, such as illegal logging, are tackled by annual inspections, surprise visits to forestry operations, and highway checkpoints for logging trucks to check their permits, among other things. As the contracts have existed for only a few years, an evaluation of the Program has not yet been undertaken (Rainforest Alliance 2001).

Case 4. Ecuador: Watershed conservation fund for Quito

Background

Ecuador's capital, Quito, receives its water from the Andean mountain range, in particular from the Cayambe-Coca and Antisana Ecological Reserves, which are inhabited by about 27,000 people. Both areas are used for agriculture and livestock grazing, which threaten the watersheds and, subsequently, the quality and quantity of water available for drinking, irrigation, and power generation downstream. The ecological reserves came under pressure in the 1970s, when petroleum development resulted in significant migration to the valley. In the 1990s, a highway was built through one of the reserves and an irrigation project was developed (Troya and Curtis 1998).

Participants in the scheme

Participants in the scheme are the municipality of Quito and private and state conservation organizations, on the one hand, and hydroelectric companies and the water users of Quito, on the other. While the latter pay for the environmental services, the municipality and its partners collect the money and either undertake compensation measures themselves or pay upstream land owners – the third party in the scheme – for changing land-use practices (Tognetti 2001).

The scheme itself: Legal and economic procedures

In 1999, the city and the conservation organizations created a fund that collects a water-consumption fee from the water users to support environment-friendly land-use practices and reforestation in the ecological reserves upstream. The goals of the programme are to maintain stream flow and water quality and to protect biodiversity by a change in land-use practices (Troya and Curtis 1998). The fund is managed by an asset-management company; decisions are made by a board of directors, which is made up of representatives of the creators of the fund and private and public users of the watershed (Tognetti 2001).

As the fee amounts were calculated based on the costs of patrolling the reserve in the first place, only 1 per cent of the revenue from hydropower generation and water-use fees goes into the fund. Until today, the small sum has been used to maintain the upstream Cayambe–Coca and Antisana Ecological Reserves. However, it is planned to expand the programme to the rest of the Condor Biosphere reserve and to determine the actual costs of water protection (Tognetti 2001).

Case 5. France: Perrier Vittel's payments for water quality

Background

In the 1980s, water quality in the Rhine–Meuse watershed in north-eastern France was threatened by the intensive agricultural practices of local farmers. Thus, companies relying on clean water for their business – namely, a bottler of natural mineral water in the area – had to choose between the cost of building filtration plants, or continuously moving on to new water sources, or of investment in the protection of current water sources; they opted for the latter.

Participants in the scheme

Participants are Perrier Vittel, the world's largest bottler of natural mineral water, which compensates about 40 dairy farmers with over 10,000 ha each in a sub-basin of the Rhine–Meuse watershed.

The scheme itself: Legal and economic procedures

In the early 1990s, Perrier Vittel decided that the protection of the water resources was the most cost-effective option and negotiated contracts with the farmers to reduce nutrient runoff and the use of pesticides. The contracts are almost purely private agreements. State institutions pay only a small percentage of total expenses, with the French National Agronomic Institute covering 20 per cent of the research costs and the French Water Agencies paying 30 per cent of the expenses for building and monitoring the use of modern barns. No formal partnership between the private and public sector was established.

Perrier Vittel pays the farmers for less-intensive pasture-based dairy farming and improved animal-waste management, and for the elimination of corn cultivation and agrochemicals. The company's intention is to reduce nitrates and pesticides and to restore the natural water-purification functions of the soil. Vittel pays unusually high compensation for an unusually long time (18–30-year contracts), "compensating farmers for the risk and the reduced profitability associated with the transition to the new technology" (Perrot-Maître and Davis 2001). Each farm received about US$230/ha/year for 7 years and Vittel spent about US$155,000 per farm on agricultural investment. It also provides technical assistance and pays for new farm equipment which, in exchange, is owned by Vittel for the contract period. Over the first 7 years, Vittel paid out about US$24.5 million for the programme. When Vittel purchased Perrier, the model was transferred to springs in southern France. Other French bottlers are now considering adopting the model (Perrot-Maître and Davis 2001).

Case 6. Philippines: Makiling Forest Reserve

Background

The Makiling Forest Reserve (MFR), 100 km south of Manila, represents an important watershed for private and industrial downstream users. The area also attracts a large number of holiday-makers and is home to about 250 households and 1,000 farmers. More than half of its area is still forested and the soil is fertile. In the 1990s, water flows decreased and water quality deteriorated in some areas. In order to face these problems, the Mount Makiling Conservation and Development Program was developed.

Participants in the scheme

Participants are local resource users and a multi-sectoral MFR Watershed Management Council that is to implement the extensive MFR Master Plan. Until 1999, the reserve was managed by the University of the Philippines, Los Baños

(UPLB), whose Vice-Chancellor for Community Affairs now serves as Chairman. Council members come from the UPLB College of Forestry and Natural Resources and the Makiling Centre for Mountain Ecosystems, as well as from sectoral user groups. Hitherto, revenues for the fund have been collected only from tourists and other users of the recreation facilities; however, water users will also be charged in future.

The scheme itself: Legal and economic procedures

The programme includes, for example, higher entrance fees to the botanical gardens and newly introduced fees to major sites of the reserve. As part of the overall strategy, local water users have agreed to pay an additional water-usage fee of US$0.014/m^3 to help finance watershed protection activities. This level of the watershed-management and protection fee was established after conducting a willingness-to-pay survey among farmers and private households in the area: according to this survey, the water users would have agreed to pay an even higher water fee. In addition to the fee, electric power generators provided seedlings for upstream reforestation efforts. The research necessary to develop the programme was partly financed by UNEP (Francisco et al. 1999).

In contrast to the other examples, no upstream household is compensated for its service. Conservation activities are conducted by the Watershed Management Council, and forest users are restricted by fees: for example, they have to pay for the gathering of forest products.

To date, this ambitious programme has not been very successful. In particular, the implementation of the watershed-protection fee has been delayed, owing to a pending court case investigating whether the university has the right to collect fees. Although the water districts are willing to cooperate in collecting the fee, low support from the university's top management and insufficient time resources for the academic initiators of the project have slowed the process. Finally, the process stalled, which was also due to the lack of ongoing financing through UNEP. However, some in-kind contributions of water users were recorded: for example, one water district provided support to reforest an area in the watershed, and resort owners volunteered to employ children of the mountain forest occupants, with the university training them.

In contrast to the water fee, the pricing of the recreation facilities has been implemented successfully. Over the last two years, UPLB has doubled the amount of fees collected (Francisco H, pers. comm., 2002).

Case 7. USA: New York City pays upstream farmers for protecting its drinking water

Background

New York City (NYC) obtains 90 per cent of its drinking water from the mostly rural Catskill/Delaware watersheds, about 200 km away. Some 77,000 people live in the area and there are some 350 (mostly dairy) farms. In 1989, a new law came into effect, according to which either drinking water had to be filtered or a watershed-control programme had to be established to minimize microbial contamination. A new filtration plant would have cost the city US$7–9 billion, including operating costs for 10 years.

Participants in the scheme

In order to avoid the costs of the new filtration plant, in 1992 NYC entered into an agreement with the watershed's farmers, forestry landowners, and timber companies. Although participation is voluntary, by December 1999 more than 85 per cent of the farmers were participating in the Program and had received money from the City (Walter and Walter 1999; Perrot-Maître and Davis 2001). In 1993, the partners created the non-profit Watershed Agricultural Council (WAC), which was to provide leadership for the improvement of land-use practices and to foster local economic development. Members include farmers and other local leaders, with representation from the NYC Department of Environmental Protection and other state and local interests (Perrot-Maître and Davis 2001).

The scheme itself: Legal and economic procedures

The 1992 Watershed Agricultural Program (WAP), which is financed completely by NYC (Hoffman 1999), is managed through the local WAC. The investment of US$1–1.5 billion over 10 years has been financed by a 9 per cent tax increase on NYC residents' water bills over a five-year period. The fund is supplemented from NYC bonds, the federal government, the State of New York, and local governments in the catchment area. Among other projects, the money is used for research, the development of Whole Farm Plans, and the implementation of best-management practices. For example, dairy farmers and foresters who adopted best-management practices were compensated with US$40 million, which covered all their additional costs. Foresters who improved their management practices (by such means as low-impact logging) received additional logging permits for new areas, and forest landowners owning 50 acres or more and agreeing to commit to a ten-year forest-management plan are entitled to an 80 per cent reduction in local property tax. NYC also paid US$472 million to improve and rehabilitate city-owned sewage-treatment plants, water-supply facilities, and dams (Perrot-Maître and Davis 2001).

According to the WAC, the Program has been an overall success so far. It represents a model in conflict resolution and watershed management (NYC Watershed Agricultural Council 2003).

Case 8. USA: Payments to farmers for the retirement of sensitive land

Background

Before 1985, public awareness of the impacts of agricultural soil erosion and water runoff of nutrients and chemicals on water resources was growing; furthermore, farm incomes were in sharp decline.

Participants in the scheme

The US Department of Agriculture (USDA) compensates farmers who are willing to retire sensitive land and "to plant long-term resource-conserving covers to improve soil, water, and wildlife resources" (Farm Service Agency 2002).

The scheme itself: Legal and economic procedures

The voluntary Conservation Reserve Program (CRP) was established nationwide in 1985 by the USDA. Under the CRP, farmers are paid to retire sensitive land from agricultural use for 10–15 years and to implement conservation practices. Originally, the Program was set up to control soil erosion; however, it now includes the protection of wildlife habitat and water quality, and the restoration of wetlands. In addition, lands located in a conservation priority area can be retired (Perrot-Maître and Davis 2001). Although the Program mainly serves lowland farmers, there are a few provisions relevant to mountain areas: for example, cropland with a high erosion index and areas suitable for the planting of living snow fences are eligible for placement in the CRP (Farm Service Agency 2002).

On average, farmers receive US$125/ha/year, based on the relative soil productivity within each county and a three-year average of the local dry-land cash rent. Furthermore, CRP covers 50 per cent of farmers' costs to establish approved conservation practices, provided that they commit themselves to the restoration of degraded wetlands and associated upland habitat for at least 10 years. Altogether, this comprises a total cost to the government of US$1.8 billion/year (Perrot-Maître and Davis 2001).

Evaluation

Synopsis of existing cases

Although the overall number of compensation schemes for environmental services in mountain areas remains rather low worldwide, one can take courage from the fact that the existing schemes have been introduced and run successfully in diverse cultural settings. Indications are that environmental service payments are a promising tool to foster sustainable development in mountainous regions worldwide.

The cases described above have some common features (table 4.1):
- The environmental service underlying the different agreements is almost always water. Siltation is a close second, in cases where siltation of irrigation channels and soil erosion are the major issues. The FONAFIFO Program in Costa Rica also compensates upstream landowners for the mitigation of greenhouse-gas emissions, as well as for the protection of biodiversity and scenic beauty.
- Problems experienced in the lower reaches of watersheds have served as incentives for setting up schemes that compensate upstream land-

Table 4.1 Summary of the various cases

Case no.	Problems downstream	Nature of the environmental service upstream	Who pays (categories)	Who receives	Involvement of public authorities	Type of compensation	Legal set-up
1	Soil salinization	Reforestation	Downstream farmer associations	Government agency, private upstream landowners	Major involvement; public agency reforests and sells salinity-reduction credits	Yearly payments per ha reforested land for 10 years	Trading scheme
2	Water scarcity, floods, siltation of irrigation channels	Reforestation, erosion control, spring and stream protection	Downstream farmer associations	Government agency, private upstream landowners	Minimal; agency only designs management plans and distributes the money	Individual contracts	Private deal
3	Siltation of hydroelectric dams, irregular stream flow	Reforestation, sustainable forestry, forest preservation	Hydroelectric companies, government fund	Private upstream landowners	Minimal; provides framework for payments, serves as mediator, increases payments	Yearly payments per ha enrolled land for 5 years	Private deal
4	Decreasing water quality and quantity	Patrolling the reserve, change in land-use practices	Water users	Fund, private upstream landowners	Major involvement; agency collects fee and undertakes compensation measures	Individual contracts	Public payment scheme, fee

	Problem	Intervention	Buyer	Seller	Intermediary	Payment	Type
5	Decreasing quality of spring water	Reduction of nutrient runoff and the use of pesticides	Private bottler of mineral water	Upstream farmers	Non-existent	Yearly payments per ha for 18–30 years, pays for new equipment	Private deal
6	Decreasing water quality and quantity		Users of recreational facilities, water users	Fund	University plays a major role		
7	Decreasing quality of drinking water	Implementation of Whole Farm Plans and best-management practices	City and water users (tax on water bills)	Upstream farmers	Major involvement; NYC completely finances the programme	Covering of additional costs of management change, reduced property tax	Public payment scheme, tax
8	Soil erosion, decreasing water quality	Reforestation, implementation of conservation practices	Government	Farmers	Major involvement; the government completely finances the programme	Yearly payments per ha for 10–15 years	Public payment scheme

owners for the environmental services of their forests (i.e. the agreements are problem driven).
- There is usually little interaction between upstream communities and downstream water users (Echavarría 2002).
- In most cases, the expected benefits have not been evaluated; the price paid for ecological services has, rather, been set by political or budgetary considerations.

Initial typology

Payment for Environmental Services (PES) schemes can be grouped into self-organized private deals, trading schemes, and public payment schemes:
- In self-organized private deals, government involvement is minimal (mediator or supplement of payments) or non-existent, and payments are made voluntarily by the downstream partner, which is either a private company or a farmer association. These cases can be found at the (sub-)watershed level, where an agreement provides private downstream entities with water services at a lower cost than traditional treatment approaches. Examples are the cases described above in France, in the Colombian Cauca River Valley, and the FONAFIFO deals in Costa Rica. Comparing those cases, the farmer associations in the Cauca River Valley invested by far the most money into the scheme.
- Trading schemes occur where governments set either a very strict water-quality standard or a cap on total pollution emissions. In Australia, the government aimed at addressing a national problem by replanting forests and trading salinity-reduction credits to downstream farmers.
- Public payment schemes are the most common mechanism. A government entity finances upstream conservation activities or reforestation from general tax revenues or water-user fees. The money usually goes into a fund which is managed by a public–private council. Examples are the cases described above in Ecuador and NYC, with New York investing the most money into the programme (Echavarría 2002).

First lessons and recommendations

According to Johnson, White, and Perrot-Maître (2001): "Overall, there is no blueprint mechanism that fits all situations – innovative mechanisms will be site-specific, will often involve elements of different approaches, and will vary depending on the nature of the ecosystem services, the number and diversity of stakeholders, and the legal and regulatory framework in place."

Some first lessons and recommendations emerge, nevertheless:

- Although PES schemes in mountain regions do not differ greatly from similar schemes dealing with water resources in the plains, they are much more rare. An option to raise their numbers rapidly would be to *integrate mountain areas into existing comprehensive environmental payment programmes*, such as the Conservation Reserve Program in the USA. Similar programmes exist in many countries.

- Downstream water users such as farmers and hydroelectric companies have an interest in watershed protection. An existing *strong legal and regulatory framework*, such as the FONAFIFO Fund in Costa Rica, helps the setting-up of local schemes because it reduces the transaction costs of establishing and maintaining the mechanism.

- Economic instruments seem to work better in an environment of well-established links between nature-management actions and products, and with well-defined rights and responsibilities. However, those conditions are the exception rather than the rule in river-basin management. Therefore, *stakeholder participation, negotiation, and institution building are critically important*, as can be seen in the Cauca River Valley and the NYC cases. Only the integration of most water users in a constituency for watershed protection guarantees the success of the scheme (Tognetti 2001). As far as institutional engagement is concerned, one can say that it is the more important, the more complex the case and the weaker the legislation in that particular field.

- *Self-organized private deals are likely to occur when there is a strong link between land-use actions and upstream watershed services.* The water services provided have to be related to private goods, such as bottled water or agricultural products. In these cases, private companies, farmers, or households have a strong self-interest in paying upstream landholders for environmental services. For example, hydroelectric companies try to avoid too irregular a stream flow and siltation of their dams. Incentives for farmers are manifold and range from a reduction in soil salinity to less siltation of irrigation channels. Another precondition of voluntary contractual arrangements is low transaction costs (i.e. the numbers of participants and the size of the watershed are limited). Thus, private deals are more likely to occur in smaller watersheds. A problem with voluntary payments of water users is that they may decline in years of economic crises, as observed in the Colombian Cauca River Valley.

- Compensation schemes where private entities, such as companies or farmers' associations, fully finance environmental services in mountain regions are *restricted to profitable industries* (Perrier Vittel) or agricultural regions where farmers get good prices for their products (Cauca River Valley). In cases where affected people or companies are financially weaker, the public sector has to provide some funds in order to establish a compensation scheme (Energía Global).

- *In contrast to self-organized private deals, public payment schemes usually occur in larger landscape systems* where the environmental services are more complex and biophysical relationships are less predictable. Other preconditions are a high number of stakeholders and a scheme according to which payments have to be collected from a large number of participants. In other cases, government institutions might be needed to organize upstream interests. Governments who pay for upstream environmental services are interested in protecting the environment, respond to public pressure, or try to avoid even higher costs of, for example, new water-filtration plants (as in NYC). In contrast to private entities, the public sector is able to reduce high transaction costs caused by a large number of stakeholders. A problem mainly associated with public payment schemes is the free-rider phenomenon (i.e. that some enjoy the service without paying for it).

- Public–private partnerships seem to work well in this sector. In the NYC case, the private side was organized and forced the city to make concessions. The final 1997 agreement was satisfactory to both sides and farmer participation is around 90 per cent. The cooperation between the farmers' associations in the Cauca River Valley and the public Cauca Valley Corporation also represents a promising example for the efficient operation of a PES scheme, with the farmers financing it and the public authority carrying out the watershed protection measures according to management plans. Thus, *public–private partnerships are most likely to occur when the private side is well organized and the public institution involved has an interest in watershed protection.* In most cases, the latter is the Ministry for the Environment or an associated regional or local authority.

- Regarding the possibility of organizing upstream forest owners and downstream water users, *it appears much easier, for instance, to form water-user associations downstream than to organize upstream landowners*; this is due to the higher financial resources downstream and their common interest in the environmental service. On the other hand, it is more urgent and much more complicated to initiate upstream cooperation. However, the effort would be encouraging because it would enable upstream people to formulate their interests and to communicate the environmental services they have to offer, contributing considerably to mountain forest protection. When planning to encourage upstream organization, it is crucial to consider that, compared with lowland areas, the area to organize can be much greater. An example is provided by large dams that affect people in numerous sub-watersheds upstream. Public entities or NGOs play an important role as initiators of upstream organization. An example is Costa Rica, where private NGOs carry out technical studies of upstream forest owners' properties and help them with the paperwork necessary to enrol in the FONA-FIFO Program.

- *Payments for environmental services must be granted for many years* in order to guarantee a long-term change to sustainable land uses and agricultural practices. Ideally, the contract states that the upstream partner has to manage his resources sustainably for a certain time, even after the payments will have ended – as in the FONAFIFO agreements in Costa Rica (Rainforest Alliance 2001). Otherwise, farmers might be tempted to clearcut their forests after they stopped receiving compensation for their services.

Pointers for starting new PES initiatives

Parameters

A number of elements are needed for developing new PES initiatives:
- It will be crucial to the success of any initiative that the resource to be protected is scarce and declining, and that its decline directly affects downstream investments or beneficiaries. This increases the likelihood that the party that could, potentially, pay compensation is likely to recognize its stake and to see benefits in entering an agreement.
- Compensation must also be high enough to serve as an incentive to upstream forest landowners to change their land-use practices. This is a complex process, in which not only individual farmers but also communities collectively must change their way of life. Compensation levels should be based on the estimated value or the economic importance of the service (Rainforest Alliance 2001).
- In many cases, education and assistance are required to enable upland farmers and communities to change their land-use patterns. Existing laws and customs have to be taken into account, for they determine rights and responsibilities, and key stakeholders need to be involved in the planning process at an early stage.
- While implementing a long-term PES scheme, major assumptions should be monitored and tested and, if necessary, adjusted or revised completely.
- The financial mechanisms chosen should fit existing institutional pa-

rameters and local customs. Great care should be taken not to introduce divisive financing schemes, which could harm equity and peace
among involved mountain communities.
- The choice of financial mechanisms will mirror regional institutional
particularities. In areas with weak public institutions, self-organized
private deals are probably the most effective; on the other hand, in
areas with strong public institutions, trading or public payment schemes
are more likely to be successful.

Initial questions

Before considering setting up new initiatives, the following questions
must be answered:
- *What ecosystem services are provided?* It is important to identify those
services that provide direct benefits to people. Furthermore, it must be
determined whether different management of the mountain environment will result in, for example, less soil erosion or higher water quality.
- *Can these services be measured and monitored?* In most mountain
regions, there are few data on the ecosystem services provided by
upstream forests. Thus, measurements and relationships from similar
regions can provide important arguments in negotiations between fi-
nancers and providers of the service.
- *What are the rights and responsibilities for resource use and management?* Knowledge about the legal/formal and the customary/informal
distribution of rights and responsibilities in a watershed is critical to the
successful introduction of market mechanisms.
- *Who supplies and who receives the ecosystem service?* A precondition
for establishing a PES scheme is to learn who owns or manages the
mountain areas that provide the service. On the other hand, there must
be people who directly profit from an enhanced ecosystem service in
order to use market tools successfully.
- *Are potential participants of the scheme aware of the environmental
problem?* After finding out about potential beneficiaries and suppliers
of the environmental service, it is important to investigate whether they
are aware of the problem. If this is not the case, measures have to be
taken in order to put the problem on the local political agenda or, if
only a few parties are involved, to raise their awareness of the problem.
- *How can downstream interests be organized? How can upstream interests be organized?* The organization of downstream interests is relatively easy. If user organizations do not already exist, the downstream
beneficiaries of the environmental service in question need to be supported with knowledge and, possibly, money in order to organize
themselves. Numerous organizations from other regions can serve as

examples. As far as upstream interests are concerned, organization is much more difficult. Here, it would be important to initiate communication in the first place, because mountain communities are often quite isolated. Furthermore, assistance in formulating their interests and offering the environmental services of their land will be necessary.

- *What is the value of the ecosystem service?* Ecosystem services that benefit people must have some economic value. However, as their real economic value is very difficult to determine, in most cases they are roughly estimated. Methods to do this are either to value the costs of replacing the service (e.g. NYC), or to value the economic activities directly depending on it (e.g. Energía Global de Costa Rica), or to conduct a willingness-to-pay survey (e.g. the Philippines). Finding the right price will be the result of negotiations between the parties involved.

- *Are beneficiaries willing and able to pay for the ecosystem service? Are suppliers willing and able to provide it?* These are the most important questions. Although one never knows if beneficiaries are willing to pay until someone makes an offer, the chances are good when the ecosystem service is scarce or declining, the economic activity linked to it is relatively important, and substitutes are expensive or unavailable. Furthermore, the beneficiaries must be convinced that the money spent is actually used for the environmental service they are paying for. In most cases, potential suppliers will provide the ecosystem service only when they are paid as much, or more, for providing it as they could obtain from alternative uses. However, especially in developed countries, potential suppliers might even offer the service for the coverage of their expenses, because they are interested in protecting the environment.

- *Is the government or an environmental NGO interested in implementing PES schemes?* The debate to introduce market tools to maintain or enhance environmental services is often initiated by governments or NGOs who bring users and providers of the service together in the first place. Thus, it might be crucial for a successful scheme to seek their support.

- *What transaction costs are involved?* Assessment of the potential for a PES scheme must recognize the transaction costs arising from stakeholder participation, negotiation or research, and monitoring and enforcement expenses. Negotiating with associations of water users or forest owners, rather than with individual water users or upstream landowners, can reduce transaction costs considerably. Governments or donor agencies might also be willing to pay such costs if the overall concept is promising (Rainforest Alliance 2001).

- *Which PES scheme is most suitable in the given situation?* In most cases, the decision will have to be made between purely private deals and public payment schemes because trading schemes seem only to be an option for industrialized countries with strong legislation. In order to decide between the two, one has to contemplate the number of stakeholders and the size of the watershed, as well as the number of people that will finance the programme. Furthermore, the validity of the beneficiaries and the necessity to organize upstream interests are of importance.

REFERENCES

Byers, E. 1995. "Environmentally sustainable and equitable development opportunities." In: Mountain Agenda, *Conserving diversity in mountain environments: Biological and cultural approaches*. Berne: Centre for Development and Environment.

Coram, J. 1998. *National Classification of Catchments for land and river salinity control. A catalogue of groundwater systems responsible for dryland salinity in Australia*. Report for the Rural Industries Research and Development Corporation, compiled by the Australian Geological Survey Organisation. Publication no. 98/78.

Echavarría, M. 2002. *Water user associations in the Cauca Valley, Colombia: A voluntary mechanism to promote upstream–downstream cooperation in the protection of rural watersheds*. Rome: UNFAO.

Farm Service Agency. 2002. Conservation Reserve Programme [http://www.fsa.usda.gov/pas/publications/facts/html/crp99.htm].

Francisco, H. et al. 1999. *Economic instruments for the sustainable management of natural resources: A case study on the Philippines' Forestry Sector*. New York and Geneva: UNEP.

FUNDECOR. 2001. *Financial technologies: Environmental services in the private-sector projects of hydroelectric development* [http://www.fundecor.or.cr/tecnologias/financieras/proyectos_en.shtml].

Hamilton, L.S., and L.A. Bruijnzeel. 1997. "Mountain watersheds: Integrating water, soils, gravity, vegetation and people." In: B. Messerli and J.D. Ives (eds) *Mountains of the world: A global priority*. London: Parthenon.

Hamilton, L.S., D.A. Gilmour, and D.S. Cassells. 1997. "Montane forests and forestry." In: B. Messerli and J.D. Ives (eds) *Mountains of the world: A global priority*. London: Parthenon.

Hoffman, R. 1999. "The New York City Watershed Agreement." *Water Resources IMPACT* Vol. 1, No. 5.

International Year of Mountains. 2002. *Mountain Forests* [http://www.mountains2002.org/i-forests.html].

Jodha, N.S. 1997. "Mountain agriculture." In: B. Messerli and J.D. Ives (eds) *Mountains of the world: A global priority*. London: Parthenon.

Mountain Forum. 2001. *Why mountains? Water resources* [http://www.mtnforum.org/members/water.htm].

Johnson, N., A. White, and D. Perrot-Maître. 2001. *Developing markets for water services from forests: Issues and lessons for innovators.* Washington: Forest Trends.

Kapos, V., J. Rhind, M. Edwards, M.F. Price, and C. Ravilious. 2000. "Developing a map of the world's mountain forests." In: M.F. Price and N. Butt (eds) *Forests in sustainable mountain development: A state-of-knowledge report for 2000.* Wallingford: CAB International.

NYC Watershed Agricultural Council. 2003. *History* [http://www.nycwatershed.org/index_wachistory.html].

Perrot-Maître, D., and P. Davis. 2001. *Case studies of markets and innovative financial mechanisms for water services from forests.* Washington: Forest Trends.

Rainforest Alliance. 2001. *Costa Rica's experimental environmental services program: Paying a fee for what forests do for free* [http://www.rainforest-alliance.org/programs/cmc/newsletter/mar01-1.html].

Rojas, M., and B. Aylward. 2002. *Cooperation between a small private hydropower producer and a conservation NGO for forest protection: The case of La Esperanza, Costa Rica.* Rome: FAO.

Tognetti, S.S. 2001. *Creating incentives for river basin management as a conservation strategy: A survey of the literature and existing initiatives.* Washington, D.C.: US World Wildlife Fund.

Troya, R., and R. Curtis. 1998. *Water: Together we can care for it! Case Study of a Watershed Conservation Fund for Quito, Ecuador* [http://www.mtnforum.org/resources/library/tncla98a.htm].

Walter, M.T., and M.F. Walter. 1999. "The New York City Watershed Agricultural Program (WAP): A model for comprehensive planning for water quality and agricultural economic viability." *Water Resources IMPACT* Vol. 1, No. 5.

5

Sustaining mountain economies: Poverty reduction and livelihood opportunities

Safdar Parvez and Stephen F. Rasmussen

Summary

There is a general consensus in mountain literature that the explanation for the observed high levels of poverty in mountain regions is to be found in the "specificities" of mountain environments – certain specific factors that are peculiar to mountain regions, such as their inaccessibility, fragility, and marginality. It is similarly argued that "positive" specificities in mountains, such as local human-adaptation mechanisms and special niches, account for the resilience of mountain communities in the wake of extreme adversity. The policy implication of this analysis is that development programmes for uplifting mountain communities will be relevant and effective only if the conceptual frameworks of such programmes internalize these specificities.

Mountain specificities are important and relevant but, on their own, do not explain why (1) on the basis of analysis of available data, socio-economic performance in mountain countries, on average, appears generally to correspond with that in non-mountain countries, and (2) for some socio-economic indicators, mountain countries appear to be performing better than non-mountain countries. It is also difficult to reconcile the mountain-specificity argument with the observed significant variations in socio-economic achievements among mountain countries themselves.

To complement this "global" analysis of comparative growth and development performance between mountain and non-mountain countries, comparisons are presented between socio-economic achievements in mountain regions and the national level for South Asian countries and China, for which such disaggregated intra-country data were available. The analysis reveals that development performance in these individual mountain regions is also varied, with some of them demonstrating stronger relative performance leading to a catching-up with their respective national socio-economic averages, although absolute levels of poverty are still higher in most mountain regions than in related lowland regions. Again, in not being able to explain the varied development performance of these mountain regions, the inadequacy of the specificities thesis becomes evident.

Even though the dearth of spatially disaggregated data means that a more robust global analysis could not be made, three important points emerge from this preliminary analysis:

- First, it is suggested that the explanation for the differentiated growth and development performance in mountain regions has broader national dimensions and cannot be fully understood or described in terms of its causal links with endogenous specificities alone. It is clear that mountain regions that have demonstrated stronger growth performance are associated with national economies that have also displayed steady growth trends.

- Second, there are several existing mechanisms through which an enabling national economic context becomes vital to supporting development of mountain regions: these include the fiscal space available to a growing national economy to channel resources and subsidies to mountain regions; the market demand that a vibrant private sector operating at the national level could generate for mountain products; the capacity of the national market to absorb surplus mountain labour; and the general diversification of mountain economies to off-farm sectors that could be promoted to relieve pressure on the indigenous natural-resource base of mountain regions. Enhanced mountain-specific investments in sectors such as infrastructure, natural-resource management, and social development continue to be critically important, but their full efficacy and potential cannot be realized in isolation from the broader national-development scene.

- Third, it is proposed that the application of a sustainable-livelihoods approach – an approach that is "people-centred" and focuses on opportunities – would provide a comprehensive framework to understand, analyse, and assess mountain-development issues and suggest appropriate policy actions.

Introduction

This chapter presents an assessment and analysis of poverty and livelihood issues pertaining to mountain people and communities. It takes account of the global context, based on available sources of information, but is also rooted in the practical experiences of working to improve the lives of people living in poor, isolated, mountain communities. The dominant perspective comes out of the experience of communities seeking to improve their livelihood opportunities.

There appears to be a general consensus on the nature of mountain-poverty issues, although caution is advised in the application of generalizations, given the variability of mountain conditions (Ives, Messerli, and Spiess 1997). In so far as livelihoods are concerned, the main problem in producing evidence to support useful global generalizations is the glaring lack of poverty data and analysis specific to most mountainous regions (Kreutzmann 2001). This is primarily due to the established practice of aggregation of socio-economic data reported in national statistics in most countries. Even where regional disaggregated data are available, mountain regions are hardly ever reported upon separately. One reason for this could be related to the general disenfranchisement of mountain communities from the national political and economic mainstream. This omission of mountain statistics leads to what is often referred to as the "statistical invisibility" of mountain regions. As described later in this chapter, this omission hinders comparative analysis of poverty in mountain areas, from which only preliminary observations can be drawn at this time.

What has been referred to above is related to the "supply"-side explanation of the stark data gaps for mountain regions. There is, however, an equally important and somewhat disturbing "demand"-side problem as well. This demand-side dimension is related to the existing consensus on mountain poverty mentioned above, according to which, mountain regions have high levels of poverty and, within their respective national contexts, mountain communities are amongst the poorest. It is surprising that this consensus continues to go largely unchallenged, despite the absence of poverty data specific to mountain regions. Even if it is accepted that general empirical and anecdotal observations provide a sufficient and reasonable basis for this consensus to hold, additional disaggregated data for mountain regions are required to assess *trends* in mountain poverty and to ascertain whether livelihoods in mountain regions have improved or worsened over time.

For a long time the dominant accepted wisdom has been to take a "resource perspective" instead of a "people's perspective" towards the analysis of mountain livelihoods. This has had implications on the kind of data that have been demanded and subsequently made available for policy purposes. This resource bias to some extent reflects the classic, well-documented inequities in highland–lowland interaction (Jodha 2000), under which the lowlands have an obvious and clear priority interest in mountain resources for purposes of exploitation and much less inclination and reason for attention to the plight of mountain people. This bias in analysis was also spurred by the relatively early interest of geographers and environmentalists in mountain regions, which, understandably, often excluded a more specific "people-centred" research perspective. This is illustrated in the concentration of attention on the study of the environment–poverty interaction in much of the research. Poverty has traditionally been seen as a cause of environmental degradation and, consequently, its elimination has been deemed a requirement for environmental protection – this is a "resource perspective." This debate has since moved on to question the extent to which poverty is correlated with environmental degradation, which is now viewed as a product of multiple factors and processes, some of which could be exogenous to mountain regions. In contrast, a "people's perspective" would highlight the importance of poverty as a consequence or impact of environmental degradation and would demonstrate how livelihood realities change when environmental resources are degraded. Work on this latter set of issues is mainly described in the sustainable-development literature but continues to be less explored and analysed. The purpose of the discussion here, however, is not to establish a (false) contradiction between concerns about environment and poverty but, rather, to note the difference in perspective that could arise from taking different approaches to the same problem.

A more systematic discussion on developing an understanding of mountain-livelihood issues was initiated with the publication of *Mountains of the World* (Messerli and Ives 1997), the first section of which comprised a series of papers on the "human dimension of mountain development." The international conference on "Growth, Poverty Alleviation, and Sustainable Resource Management in the Mountain Areas of South Asia" (Banskota, Papola, and Richter 2000) was a further comprehensive attempt to analyse livelihood issues from a regional perspective and, in the process, made available considerable socio-economic data on mountainous regions of the South Asian countries. This chapter draws much from such valuable sources, while at the same time taking the perspective of development practitioners immersed in working in partnership with mountain communities.

Mountain livelihoods: Inferences from global experience

The broad conclusions about mountain poverty and livelihood issues and dynamics may be summarized as follows.

Acute levels of poverty

There is a general consensus that mountain regions are characterized by acute levels of poverty. The discussion is summed up in the shape of the following "facts" about mountain poverty (Ives 1997): (a) mountainous/landlocked countries are often among the poorest in the world (for example Nepal, Ethiopia, Bolivia, and others); (b) mountain regions in less-developed countries are generally among the poorest regions in these countries (for example, Uttar Pradesh in India, Yunnan and Xinziang in China, northern Pakistan); and (c) mountain regions – even in developed economies, such as Austria and Switzerland – are comparatively less affluent than the lowland areas of these countries. This generalization receives further examination later in this chapter, especially in the light of findings that suggest that it is not always evident.

Specificity of mountain environments

It is argued that there are structural mountain-specific reasons for the persistence of high poverty levels, particularly features such as the inaccessibility, fragility, and marginality of mountain environments. At the same time it is recognized that the wide diversity of conditions across mountain regions makes it difficult to apply uniform poverty-reduction strategies, and it is suggested that differentiated mitigation activities should be pursued while addressing these shared characteristics. There is also recognition of specific "natural niches" as well as "human adaptation mechanisms" that contribute towards sustaining and improving mountain livelihoods. The derived policy implication of mountain specificities is that these have to be factored in as central considerations in designing and implementing mountain-development programmes and initiatives, without which such programmes are likely to fail.

Trade-off between greater integration and more dependence

A strategy that is often used to help mountain regions to overcome the constraints arising from the adverse specificities mentioned above has been to seek greater physical and market integration with the lowlands. This is made possible through substantial national investments into constructing transportation and communication networks to facilitate highland–lowland interaction and to enhance access of mountain regions to major markets. More recently, some of the potential implications of globalization and movement towards free trade on mountain communities have begun to be analysed and appreciated. It is in these contexts that the centre–periphery dichotomies and concepts have been extended to explain the inequities of exchange in highland–lowland interactions (Grötzbach and Stadel 1997; Jodha 2000). For example, it is argued that greater integration has made local farmers more dependent on the vagaries of national agrarian policies conceived primarily from the perspective of benefiting the lowlands; that "extractive" policies of national governments have led to a resource drain from the highlands to the lowlands without adequate compensatory mechanisms; that mountain tourism has had a negative impact on the environment and culture of the mountains and that mountain people do not substantially benefit from tourism-related revenues and resource flows; and that globalization would further marginalize poor mountain communities. Policy implications of this analysis from a livelihoods perspective suggest that, while greater lowland–highland integration is a given, conditions must be created through "partnership and sharing of gains" as well as by establishing "compensating mechanisms" that ensure that mountain communities also benefit equitably from the resulting resource and market flows (Jodha 2000).

Need for long-term state support and subsidy

There is general consensus that mountain regions are in need of long-term support from the state to compensate for the structural disadvantages caused by the specificities mentioned above. The example of mountain regions in most developed countries (for example, the Alpine regions of Europe) shows that they have prospered in recent times under high levels of institutionalized state protection and subsidy. The underlying assumption is that the sustainability of livelihoods of mountain communities is contingent on long-term public support, even in the presence of growing private highland–lowland interactions, given that the exchange inequities involved do not favour mountain communities. For the mountain people of developing countries, the situation clearly is even more precarious and the call on government support is even more urgent and justified. This is because of generally weak national economic conditions with stifled private-sector growth, providing relatively little opportunity for economic diversification in mountain regions, fewer opportunities for intensification of natural-resource use, and consequent widespread migration to the lowlands. It is pertinent to mention here that the strategic importance and conflict-zone status of many mountain regions, instead of the economic and livelihood imperatives of the local communities, has often been the compelling factor for the governments to invest, particularly in infrastructure (Kreutzmann 2000).

Poverty and environmental degradation

The first section of this chapter mentioned the emphasis that has often been put on the poverty–environment nexus. This began from a focus on the environment itself and on how poverty contributed towards degradation of natural resources; it later diversified to a concern for sustainable development, outlining the centrality of the natural-resource base as a livelihood provider for millions of poor mountain people and the consequent need for its conservation and protection. Later on, the finding that it is not poverty alone that adversely affects the environment challenged the cause-and-effect relationship theory between poverty and environmental degradation. Additional exogenous factors undermine mountain environments, such as the impact of lowland-sponsored extractive industries and environmentally unfriendly policies of the state (Prakash 1997).

Diversification, migration, and changing gender roles

Rapid population-growth rates have forced mountain communities to adopt livelihood-diversification strategies to supplement incomes derived from traditional farm sources. Opportunities for diversification have mainly occurred through two means – growth of mountain tourism and out-migration. Tourism has quickly become the major off-farm source of livelihood in many communities – catalysed, on the supply side, by easier access to mountains made possible by investments in roads and other infrastructure and, on the demand side, by the enhanced standard of living of urban people, allowing them more substitution possibilities between work and leisure. At the same time, there has been a growing recognition and concern for the implications of tourism on the social and cultural aspects of mountain life (Godde, Price, and Zimmermann 2000).

Migration to the lowlands has provided a much-needed outlet for excess mountain labour and manpower. It is estimated, for example, that migration rates for adult males in the mountainous regions of South Asian countries have been in excess of 40 per cent (Papola 2001). Such high rates of migration have led to a significant change in gender roles in the region, with women forced to take on added roles for both farm and domestic household work, as a result of which their workloads have greatly increased (Ives 1997) and the number of female-headed households in many mountain communities has increased significantly. The recent historical trajectory of mountain regions in many developed countries also has been that improved education opportunities for mountain people, combined with the relatively better job opportunities of the lowlands, have resulted in significant out-migration. Along with factors such as greater highland–lowland economic interaction and increased state subsidies to mountain regions, some of the effects of migration (such as lower population levels in the mountains and monetary remittances from family members working in the lowlands) have also contributed to the improvement of local conditions. However, some commentators have pointed out that the flip side of out-migration has been a declining population density in the highlands of developed countries, which makes people's lives difficult and reinforces poverty for the less able.

The purpose of this section has been to briefly introduce the main focal points of the mountain-livelihoods discussion based on published literature. It is, clearly, not possible to conduct a detailed analysis of these points within the confines of this chapter, which, instead, focuses on the following:
- an empirical reassessment of the evidence that is available on mountain poverty and livelihoods;
- an examination of how trends in the existing data might be explained; and
- a proposal of a way forward towards improving the analysis necessary to lead to better policies for the development of mountain communities.

On topics such as the poverty–environment link or changing gender roles, it is not possible to add much to what is already known, in the absence of additional empirical data and information.

A reassessment of mountain poverty

For the purposes of this chapter, the foregoing discussion might be concluded by saying that global experience appears to establish that there are high levels of poverty in the mountainous regions of the world, and that these poverty levels are attributable mainly to the specificities of the mountain environments that have to be taken into account while setting policy priorities for reducing poverty. However, there are data (albeit limited to a few countries and incomplete in terms of being spatially disaggregated) that put mountain-poverty issues within the overall perspective of global and national poverty trends. Such a preliminary analysis indicates that, despite the persistence of mountain specificities and the accompanying need to devise targeted strategies, the long-term growth and poverty-reduction prospects in mountain regions are inextricably linked to the growth and expansion of their respective national economies.

This analysis looks at a comparative assessment of socio-economic attainment of mountain regions compared with non-mountain regions and is conducted at two levels. First, selected socio-economic variables for mountain countries are compared with the respective averages of these variables for non-mountain countries to see if there are achievement differences between the two comparison categories. Recognizing that the definition of mountains is problematic (Ives, Messerli, and Spiess 1997) – and that, correspondingly, the definition of mountain countries is more so – mountain countries are defined as those that have the majority (more than 50 per cent) of their land area covered by mountains, following the classification developed by the Swiss Agency for Development and Co-operation (2001) (appendix 5.1). There are 53 such countries, although data could be found for only some of them. Second, comparisons between socio-economic achievement in mountainous regions and the national level for some South Asian countries and China are assessed. Much of these data became available through the published papers coming out of a conference organized by ICIMOD in 2000 (Banskota, Papola, and Richter 2000). As mentioned earlier, similar data for mountain regions in other developing countries appear to be more limited: although a considerable effort was made to find similar data from other mountain regions, it appears that this is not available easily, if at all. The addition of such data in future would add much to the preliminary analysis in this chapter.

Global comparisons

Table 5.1 compares various socio-economic variables of mountain countries with those of non-mountain countries and shows that, on average, mountain countries appear to have higher achievement rates than non-mountain countries: average income per capita in the mountain countries is 35 per cent higher than that in the non-mountain countries; infant mortality is 37 per cent lower; life expectancy for both males and females is higher; and the illiteracy rate for both adult males and females is lower.

In order to control for the fact that some mountain countries are among the most developed in the world, standard World Bank categorization of high-income, medium-income, and low-income countries (World Bank 2001) is used to show mountain–non-mountain comparisons within these different categories. It is found that, whereas gross national product (GNP) per capita in the medium-income category is lower in the mountain countries than the non-mountain average, it is actually

Table 5.1 Comparison of selected economic and social indicators

Income	Countries	Per capita GNP (1999)	Under-5 mortality rate per 1,000 persons (1998)	Life expectancy (years) at birth (1998)		Adult illiteracy rate per cent above age 15 (1998)	
				Male	Female	Male	Female
Overall	Non-mountain	5,155	73	62	66	20	31
	Mountain	6,944	46	66	71	18	27
Low	Non-mountain	380	134	52	55	32	51
	Mountain	368	90	57	62	29	39
Medium	Non-mountain	2,791	34	66	72	10	15
	Mountain	2,219	33	69	74	15	26
High	Non-mountain	23,001	6	75	81	4	10
	Mountain	23,073	6	75	81	1	2

Source: Computed from World Bank (2001).

Table 5.2 GDP growth-rate comparisons

Countries	1980–1990	1990–2000
Mountain	3.0	1.3 (2.8)[a]
Non-mountain	2.8	2.4

Source: Computed from Swiss Agency for Development and Cooperation (2001).
a. Excluding Central Asian mountain countries.

slightly higher for the mountain countries in the high-income category and it is almost the same across mountain and non-mountain low-income countries. Although there is little difference in social-sector achievements between the mountain and non-mountain countries in the high-income category (except for adult illiteracy, which is lower in mountain countries), infant mortality is lower and life expectancy higher in mountain countries in both low- and medium-income categories. All this would indicate that, from the point of view of economic and social-achievement indicators, a general characterization of mountain countries as being necessarily distinct in the global perspective is not warranted.

Table 5.2 shows GDP growth-rate trends in mountain countries compared with respective non-mountain averages. The average annual mountain growth rate for the period 1980–1990 is estimated to be 3.0 per cent – slightly higher than the non-mountain average of 2.8 per cent. For 1990–2000, the growth rate of mountain countries was 1.3 per cent compared with a non-mountain average of 2.4 per cent, although it is important to mention that the mountain countries in the sample included several Central Asian countries that came into being after the disintegration of the Soviet Union and which went through severe economic crises and instability in this period. Growth rates for countries such as Tajikistan, Kyrgyzstan, Georgia, and Azerbaijan were in the range of −7.5 to −10 per cent. Excluding these countries, the average mountain growth rate comes to 2.8 per cent for this period – again, higher than the non-mountain average.

Table 5.3 presents a comparison of human development indices (HDIs) (UNDP 2001) between mountain and non-mountain countries for the three categories of high, medium, and low human development. Average HDIs for high- and low-income mountain countries are higher than the corresponding averages for non-mountain countries, whereas the average HDI for medium-income mountain countries is less than that of the average for mountain countries. Nevertheless, the overall average HDI rank for mountain countries is 80, which is right in the middle of the total ranking of 160 countries.

Table 5.3 Human Development Index comparisons

Countries	Human Development Index		
	High	Medium	Low
Mountain	0.892	0.652	0.433
Non-mountain	0.875	0.693	0.403

Source: Computed from UNDP (2001).

From this analysis it could be concluded that average socio-economic achievements and rates of economic growth in mountain countries generally correspond with respective non-mountain averages and, for some indicators, mountain countries appear to have performed better. Even taking into account that this is necessarily a tentative analysis because of the general lack of spatially disaggregated data, it is not easy to reconcile this preliminary finding with the generally accepted description of adverse mountain specificities leading to lower achievement levels. A related concern about the specificity argument is that adverse specificities should affect all mountain countries more or less uniformly, whereas the observed variation in socio-economic achievements by mountain countries is considerable. Put in a slightly different way, the fact that some countries have been more successful than others in overcoming negative specificities such as marginality, inaccessibility, and fragility, or have made better use of positive specificities such as abundance of natural resources and responsive local human-adaptation mechanisms, needs further explanation. In summary, (a) socio-economic achievements in mountain countries generally correspond to the respective non-mountain averages despite adverse specificities, and (b) within mountain countries there is evidence of high, medium, and low performance, implying the possible presence of influences additional to mountain specificities.

Intra-country comparisons from South Asia and China

As already mentioned, some socio-economic data comparisons between mountainous and non-mountainous regions for South Asian countries and China have recently become available. These data are summarized and assessed below to analyse poverty and growth in mountain regions within their respective national perspectives.

Bangladesh (Shelley 2000)

Despite its long period of insurgency, the Chittagong Hill Tracts region, inhabited by about a million people, has some socio-economic indicators that are actually above the national averages for Bangladesh. According to the 1991 census, the literacy rate (having historically lagged behind the rest of the country) had become 28 per cent compared with the national average of 25 per cent, the labour force participation rate was 35 per cent as against 27 per cent nationally, and adoption of modern rice varieties stood at 66 per cent compared with the national average of 45 per cent. In the absence of any direct income or poverty comparisons between this mountain region and national levels, it is not possible to assess definitively whether levels of poverty were higher or lower, but achievements on these three indicators make it doubtful that poverty levels could have been higher in the Chittagong Hill Tracts region.

Bhutan (Lhamu, Rhodes, and Rai 2000)

Bhutan is a mainly mountainous country and available data pertain only to the country as a whole. Bhutan is reported to have made significant progress both economically (GDP per capita increased from US$100 in 1977 to US$551 in 1999), as well as in terms of human development (its HDI ranking of 0.510 places it in medium human-development countries despite its low income status). This growth has been achieved while adequately maintaining the country's natural-resource base, with agriculture continuing to be the main livelihood source of most households.

India (Joshi 2000)

Relatively more comprehensive data that allow comparisons between the mountain states and national-level achievements are available for India. Such data are summarized in table 5.4. As this table makes clear, in 1980–1981, incomes per capita of two of the eight mountain states were higher than the national average, while those for the other states were in close proximity, being in the range of 80–87 per cent of the national level. By 1996–1997, incomes per capita for another two states (Arunchal Pradesh and Mizoram) had become higher than the national income per capita and Nagaland had come very close to the national level. However, incomes in the states of Jammu and Kashmir, Tripura, Manipur, and Meghalaya had come down as a percentage of national incomes, substantially so for the first two states. Joshi points out that an important explanation of this is that three of these four states had been "in the throes of serious problems of insurgency" during this period, adversely affecting their economic growth.

Nevertheless, it is clear that, whereas average annual growth rates for most of the mountainous states are below the Indian average (which surged following structural adjustments in the national economy), they have still been quite respectable. The levels of poverty in four mountain states appear to be higher than the national poverty level although

Table 5.4 Economic data for mountain states in India

State	Proportion of per capita net domestic product of mountain states to national level (percentage)		Annual growth rate	Poverty level (percentage of population)
	(1980–1981)	(1996–1997)	(1992–1997)	(1994)
Arunchal Pradesh	96	104	4.9	39
Himachal Pradesh	105		4.3	28
Jammu and Kashmir	109	58	3.9	25
Manipur	87	65	5.2	34
Meghalaya	84	73	2.8	38
Mizoram	79	116		26
Nagaland	89	97	6.4	38
Tripura	80	47	7.3	39
India	100	100	5.6	36

Source: Constructed from various tables in Joshi (2000).

they are lower for the other four. Thus, no consistent relative mountain-poverty trend generalization emerges. Overall, data on mountain regions in India display considerable diversity, with some mountain states ahead of the national averages for economic indicators, others not far behind, and yet others showing disappointing trends, at least partly related to civil unrest.

Pakistan

A time series of disaggregated household-level income and expenditure data is available for the northern mountain regions of Pakistan, collected and analysed by the Aga Khan Rural Support Programme (Aga Khan Rural Support Programme 2000; Malik and Wood 2003). This was the only micro-level rigorous analytical work on mountain poverty and live-lihoods spread over 12 years (1991–2003) that the authors could find when writing this chapter. Tables 5.5 and 5.6 summarize the results of the comparative analysis carried out on income and poverty trends in three different regions of northern Pakistan (Gilgit, Baltistan, and Chitral) relative to the national level.

Table 5.5 shows that, although the northern mountain regions in Pakistan lag behind the national average for incomes per capita, there has been some significant catching-up in the various sub-regions relative to the national level, albeit at a differentiated rate. Thus incomes per capita in Gilgit had increased from 32 per cent of national incomes per capita in 1991 to 69 per cent by 2001, compared with an increase from 33 per cent

Table 5.5 Income comparisons for Northern Pakistan

| | Proportion of per capita income in Northern Pakistan relative to national level (percentage) | | |
	(1991)	(1997)	(2001)
Gilgit	32	62	69
Baltistan	24	49	57
Chitral	33	44	46
Pakistan	100	100	100

Source: Aga Khan Rural Support Programme (2000); Malik and Wood (2003).

to 46 per cent in Chitral during the same period. The much higher levels of public-sector development expenditure in Gilgit and Baltistan and their year-round road link with the rest of the country are largely responsible for their better economic performance relative to Chitral. Together with the high growth rate in incomes per capita, poverty declined steeply (table 5.6), again at differentiated rates in the various sub-regions: in 2001, the poverty level in Gilgit was estimated to be less than the national level, whereas poverty in Baltistan was only slightly higher than this level.

Figure 5.1 presents an interesting and useful analysis of the sensitivity of the poverty index to changes in income in northern Pakistan (Malik and Wood 2003). This analysis is helpful in highlighting the issue of socio-economic vulnerability in mountain regions, rather than only the level of absolute poverty. It shows that the level of vulnerability is high: if incomes per capita fall by 10 per cent, the poverty level in 2001 goes up to 40 per cent from 34 per cent and the poverty level would increase to 76 per cent if there were a 50 per cent decline in incomes per capita. Thus, even when the poverty level has declined substantially, the level of vul-

Table 5.6 Poverty comparisons for Northern Pakistan

| | Percentage of population below the poverty line | | |
	(1991)	(1997)	(2001)
Gilgit	62	35	29
Baltistan	76	43	34
Chitral	68	50	42
Pakistan	27[a]	26	32

Source: Aga Khan Rural Support Programme (2000); Malik and Wood (2003); Government of Pakistan (2003).
a. This figure is for 1993–1994.

Vulnerability

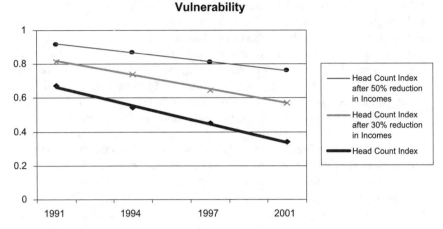

Figure 5.1 Sensitivity of Poverty Head Count Index to changes in incomes for Northern Pakistan.
Source: Malik and Wood (2003)

nerability remains high and represents a key challenge for livelihoods in the region. Vulnerability as a major livelihoods issue in mountain regions is linked to considerations of mountain specificities. This issue is taken up further in the final section of this chapter on the sustainable-livelihoods approach to understanding mountain livelihoods.

China (Ruizhen 2000)

The mountainous region of the Qinghai–Tibetan Plateau in China has somewhat lower achievement indicators than the rest of the country. Incomes in Tibet and Qinghai are 63 and 87 per cent of the national Chinese average. Average savings per person and overall level of technical skills in the plateau are also comparatively lower. Poverty levels in 1979 in Qinghai and Tibet were estimated at 30 per cent compared with the national rural average of 24.4 per cent. However, the poverty level in Tibet had come down to 19.2 per cent in 1995 as a result of sustained government interventions for poverty reduction.

Some points arising out of the discussion on mountain regions in South Asian countries and China are worth noting. First, although it is true that mountain regions have traditionally been among the poorest regions in their respective countries, and persistently high levels of poverty in mountain regions still present considerable challenges to development, there is evidence of significant *relative* growth and development, as well as poverty reduction, in many of these regions that, in some cases, has led to a catching-up with the national economies. Bhutan as a whole has made significant progress; and Arunchal Pradesh, Mizoram, and Naga-land in India, with northern Pakistan, and the Tibetan Plateau in China, have all seen significant growth and poverty reduction in their respective national contexts. Second, there are considerable variations in the rate of economic growth and poverty levels within the mountain regions of indi-vidual countries. In India, several mountain states appear to have per-formed better than other mountain states and, in Pakistan, the rate of poverty reduction in Gilgit and Baltistan regions has been higher than that in Chitral. This leads to the observation that it would be erroneous to consider mountain regions *as a whole* to be necessarily among the poorest regions in their respective countries, although there are moun-tain *sub-regions* with extreme poverty levels that would still count as being among the poorest regions in their own national contexts.

Preliminary conclusions on available data and trends

In mountain regions of the developed world, poverty is not usually an important issue, although statements about the inequities faced by mountain people, linked to mountain-specificity arguments, are similar for each of these regions. A recently completed study analysed available data for the countries of the European Union and the states likely to join it in the near future (Accession States) (Copus and Price 2002). The def-inition of mountain area was derived from a recent study that developed consistent criteria at the global scale (Kapos et al. 2000). One of the ini-tial problems identified in the analysis is that the boundaries of the areas defined as "mountain" rarely match well with the boundaries of statisti-cal reporting districts. Consequently, analyses assessed the implications of different thresholds of "mountainousness." The results were similar to those discussed above – namely, that conditions in mountain areas tended to reflect national trends. However, when two other variables were incorporated, clearer trends emerged for European countries. First, mountain areas with large towns (population over 100,000) had con-sistently higher GDP per capita than those without. Second, when a measure of "peripherality" was included into the analysis, it was found that the combination of "mountainousness" and "peripherality" was re-flected in lower GDP per capita and loss of population, whereas these trends for one variable or the other were not consistent. These findings are very preliminary, particularly because the statistical reporting areas, as in South Asia and China, are large and generally include both moun-tain and non-mountain land. Overall, findings from both the Asian and European analyses reinforce the initial conclusion, from the previous

section on global comparisons – that observed trends in mountain poverty, livelihoods, and demography are closely linked to global and national achievement levels, while also displaying great variability in both
inter- and intra-mountain contexts, and are not entirely explained by
mountain specificities and therefore are not adequately addressed by
simple policy generalizations. At the same time, it would be important to
re-examine this preliminary finding as better data for more parts of the
world become available. With the availability of more data, it might also
be useful to try to define poverty lines for mountain regions that may
differ from national-income poverty lines. This could take account of
dissimilar (and, likely, higher) caloric-intake requirements in mountain
regions. Even so, for comparative analysis of mountain-poverty levels
with national trends, it would still be necessary to use uniform poverty
lines. Finally, availability of additional data and more detailed analyses
would make possible the assessment of the robustness of levels and
trends in mountain poverty, especially with regard to the sensitivity of
mountain-poverty estimates to the use of alternate poverty lines.

Policy implications

Experience worldwide with promoting sustainable livelihoods for poor
people in mountain regions focuses on several standard recommendations. These recommendations include focusing on strategies for agricultural intensification while protecting the natural resource base, diversification of livelihood sources, improving physical access and infrastructure,
developing mechanisms to compensate the highlands for the use of the
mountain resource base by the lowlands, and the necessity to address the
inequities that prevail in highland–lowland interactions.

   These recommendations make eminent sense and should be pursued.
Experience appears to indicate, however, that such actions are not
enough to bring mountain regions to par with national levels and, in
themselves, do not provide sufficient economic opportunities and options for mountain people. The analysis in this chapter suggests that two
broader dimensions should be added to the discussion on policies and
strategies for mountain development and poverty reduction. The first is a
"national dimension" that seeks to place mountain development in the
national perspective; this dimension is traditionally understated because
of the focus on the uniqueness of mountain environments and issues.
The second is a "framework dimension" that uses a "people-centred"
framework – namely, a sustainable-livelihoods approach – for analysis
and assessment of mountain-livelihood issues and strategies.

Placing mountain development in the national perspective

The analysis in this chapter argues that mountain specificities fail to explain all of the observed variances in socio-economic achievement indicators across different mountain countries or even within mountainous regions of individual countries. Although a focus on mountain specificities is very useful in directing attention to the uniqueness of mountain environments, it must also be recognized that the explanation for this varied growth performance has broader, national dimensions. In fact, it could be said that, with growing highland–lowland interaction and economic integration, the respective national contexts of the mountain regions have particular relevance to defining the pace and scale of economic growth, and that the development of mountain regions cannot be achieved without first taking account of national (and even international) contexts. It is obvious, for example, that mountain regions in developed countries are better off than mountain regions in developing countries. More important, though, is the observation from available data that, even within developing countries, mountain regions that have grown faster are associated with the faster-growing national economies. In India, growth rates of mountain states, although relatively lower in magnitude, have tended to follow the consistent healthy growth trends in the national economy. In Nepal, poverty in the mountains has remained very high at a time when national economic growth has been stagnant and levels of poverty have been increasing. In Pakistan, data from the high mountain areas show that the growth performance of household incomes was higher in the first half of the 1990s and much lower in the second half, again following broad national trends. This is not to argue that substantive national economic growth alone would translate into better development prospects for mountain people through a classic trickle-down effect: what is being underscored is that robust national economic growth is important in providing improved livelihood options for mountain people, especially given the increased interdependency and linkages between highlands and lowlands.

Mountain regions benefit in at least four ways from a growing national economy. First, they benefit through state subsidy and resource flows: clearly, a state with more fiscal space has more spending and resource allocation flexibility for mountain regions. Second, a growing national economy creates demand and market access for highland products. Third, steady economic growth conditions absorb surplus mountain labour, making possible migration to the lowlands for people who cannot find opportunities in limited mountain economies. Finally, links with the national economy promote livelihood diversification in the mountains and can help contain the pressure on natural resources exerted by growing populations. At the same time, there are genuine concerns with other aspects of economic integration – for example, environmental concerns related to tourism, or increased workloads for mountain women as a result of male migration to the lowlands. These concerns suggest a need to develop responsive strategies that obtain greater advantages for mountain people from national policies while also incorporating adequate protection and adaptation mechanisms for mountain regions. An example of such a strategy, still not commonly practised, is environmental service contracts whereby downstream users of water compensate mountain communities to protect upstream watersheds and water sources (Koch-Weser and Kahlenborn, this volume, ch. 5).

Seeking opportunities for greater market integration and private-sector participation might also be important from another perspective. One main reason for the generally laggard state attention to mountain regions, especially in developing countries, has been their relatively very small population base (2–12 per cent of the total world population, depending on the definition used) and, consequently, their lack of political voice. Even so, direct state-resource transfers, sometimes even proportionally greater than those for other parts of the country, have been made available either on grounds of extreme poverty or for strategic military reasons, given that mountain regions in many parts of the world have long been conflict zones (Libiszewski and Bachler 1997).

Although some mountain regions are still among the poorest parts of their countries, the preliminary analysis above shows that it has become difficult to sustain the argument that mountain regions, as a whole, are necessarily always the poorest regions in their various national contexts. This could possibly result, in future, in reduced amounts of state subsidy; if so, the compensating factor could only be increased private-sector business activity and resource flows based on exploitation of comparative advantage for mutual benefit.

This discussion clearly does not detract from the fact that a supportive national and private-sector environment would still need to be complemented by specific investments in mountain regions in infrastructure, natural-resource management, and social sectors, for example. Nor does it mitigate the huge challenge of trying to reduce persistently high levels of poverty in mountain regions. What this analysis shows is that the scale, efficacy, potential, and long-term impact of such mountain-specific investments is clearly linked to national economic performance and cannot be realized or sustained outside of this overall development context.

A sustainable-livelihoods approach

There generally appears to be a certain degree of hesitation to embrace mainstream development frameworks for mountain regions, on the

grounds that such frameworks cannot adequately take mountain specificities into account (Papola 2001). However, given the need to place
mountain specificities in the broader national context (as has been argued in this chapter), it seems sensible to agree on the usefulness of applying well-established and agreed multidisciplinary frameworks and indicators to facilitate comparative assessment and analysis of issues in
mountain regions with respect to broader development trends (Kreutzmann 2001). Such an approach could also help in attracting more concerted and focused attention to the state of underdevelopment in mountain regions, given that national policy makers might be able more easily
to identify with (and grasp the significance of) such indicators. In order to
systematically apply mainstream methodologies to analysing mountain-
development issues, there is an urgent need to reduce the current "statistical invisibility" of mountains through collection and dissemination of
more organized data on social and economic issues to enable better-
quality comparative analysis and to facilitate the creation and implementation of relevant development policies and strategies.

A sustainable-livelihoods approach (Carney et al. 1999), now widely
used in development work, offers a useful framework to analyse mountain livelihood and poverty issues and to derive relevant policy implications. This is essentially a "people-centred" approach that seeks to
analyse and understand the development and dynamics of livelihood
strategies based on the assets and opportunities available within the context of the relevant external and institutional environment. The key focus
in this approach is on developing an understanding of how people and
communities make use of their existing human, social, natural, physical,
and financial capital to adopt livelihood strategies to overcome vulnerability associated with exogenous and endogenous shocks and to achieve
desired outcomes such as increased well-being, improved food security,
higher income levels, and sustainable use of natural resources. In short,
the framework "offers a way of thinking about livelihoods that helps
order complexity and makes clear the many factors that affect livelihoods" (DFID 1999). A common diagrammatic representation of the
sustainable-livelihoods analysis framework is shown in figure 5.2.

There are no well-documented applications of the sustainable-
livelihoods approach to mountain environments and it is not the mandate
of this chapter to develop such an application. Nevertheless, there are
various positive aspects for understanding mountain-development issues.
In the first place, as already mentioned, the sustainable-livelihoods
approach helps to focus attention on people and their livelihoods instead
of just on resources and their depletion. This could help to broaden the
notion of sustainable mountain development to include its important so-

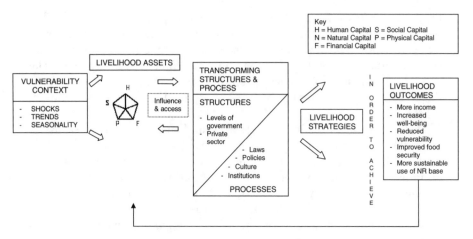

Figure 5.2 Sustainable-livelihoods framework (H, human capital; N, natural capital; F, financial capital; S, social capital; P, physical capital; NR base, natural-resource base).
Source: DFID (1999)

cial dimensions and, in so doing, could balance the traditional focus on placing natural-resource issues at the centre of the analysis of the development of mountain regions.

Secondly, the concentration of a sustainable-livelihoods analysis is on opportunities and not on constraints. This would be a welcome shift of emphasis in the mountain literature, which has traditionally paid more attention to the many obvious constraints defined as negative mountain specificities but has given less thought to the positive specificities. As a result of this concentration on negatives, there is an almost universal sense of inevitability about poor prospects for improving the livelihoods of mountain people. On the other hand, positive specificities need further elaboration and would be important to understanding the "hows and whys" of mountain-livelihood development in the face of extreme adversity. One of the most obvious omissions is the notable absence of a discussion on the generally high levels of social capital in mountain regions based on cohesive communities with well-established traditions of cooperation and collective work (Wood and Shakil 2003). This has been either overlooked or couched in rather narrow terms, such as when references have been made to the role of communities in collective management of natural resources. Social capital also manifests itself in the multifarious survival and coping strategies that communities undertake in the face of considerable vulnerability; the physical infrastructure that they are, collectively, able to build and subsequently to maintain; the business associations that emerge to take advantage of economies of scale in export and marketing of mountain products; and the local institutions that are fostered for long-term sustainable development. The recognition of the importance of social capital immediately establishes the relevance of community-based development programmes in mountain environments.

Thirdly, through defining poverty as a fundamental lack of basic assets – whether physical, natural, human, financial, or social – a sustainable-livelihoods framework has the potential to directly incorporate considerations of mountain specificity (such as isolation and fragility) that contribute to vulnerability of livelihoods in mountain environments. The analysis of vulnerability is another area for further research that might have important implications for understanding and improving livelihoods in mountain regions.

Finally, a sustainable-livelihoods approach makes possible an integrated and explicit analysis of the various conditioning factors, both exogenous and endogenous, that catalyse or impinge upon mountain-livelihood development. Thus, exogenous influences (such as the impact of the national economy and globalization), or endogenous area-specific factors (such as lack of effective women's participation and religious intolerance), could all be made part of the framework and fed into the policy-development process. In this way, a sustainable-livelihoods approach has the potential to improve the understanding of the (often relatively high) levels of socio-economic vulnerability faced by mountain communities due to adverse conditioning factors that might exacerbate the impact of negative mountain specificities.

Appendix 5.1

Countries with >75 per cent mountain area: Andorra, Armenia, Bhutan, Bosnia and Herzegovina, Georgia, Kyrgyzstan, Lebanon, Lesotho, Macedonia, Montenegro, Nepal, Reunion, Rwanda, Switzerland, Tajikistan.

Countries with 50–75 per cent mountain area: Afghanistan, Albania, Austria, Azerbaijan, Burundi, Cape Verde, Chile, China, Comoros, Costa Rica, Djibouti, El Salvador, Eritrea, Greece, Haiti, Honduras, Iceland, Iran, Italy, Japan, Laos, Morocco, New Caledonia, New Zealand, North Korea, Norway, Serbia, Slovakia, Slovenia, Swaziland, Taiwan, Turkey, Vanuatu, West Bank, Western Samoa.

Source: Swiss Agency for Development and Cooperation. 2001. *Mountains and people: An account of mountain development programmes by the SDC.* Berne, Switzerland: SDC, p. 8.

REFERENCES

Aga Khan Rural Support Programme. 2000. *An assessment of socio-economic trends and impact in Northern Pakistan (1991–1997)*. Pakistan: AKRSP.

Banskota, M., J.S. Papola, and J. Richter (eds). 2000. *Growth, poverty alleviation and sustainable resource management in the mountain areas of South Asia*. Germany: Deutsche Stiftung fur internationale Entwicklung.

Carney, D., M. Drinkwater, T. Rusinow, K. Neefjes, S. Wanmali, and N. Singh. 1999. *Livelihood approaches compared: A brief comparison of livelihood approaches of the UK DFID, CARE, Oxfam, and UNDP*. London: DFID.

Copus, A.K., and M.F. Price. 2002. "A preliminary characterisation of the mountain areas of Europe." Unpublished report to Euromontana.

Department for International Development (DFID). 1999. *Sustainable Livelihoods Guidance Sheets*. London: DFID.

Godde, P., M.F. Price, and F. Zimmermann (eds). 2000. *Tourism and development in mountain regions*. Wallingford: CAB International.

Government of Pakistan. 2003. "Draft poverty reduction strategy paper – summarized version." April.

Grötzbach, E., and C. Stadel. 1997. "Mountain Peoples and Cultures." In: B. Messerli and J.D. Ives (eds) *Mountains of the world: A global priority*. New York: Parthenon.

Ives, J.D. 1997. "Comparative inequalities – Mountain communities and mountain families." In: B. Messerli and J.D. Ives (eds) *Mountains of the world: A global priority*. New York: Parthenon.

Ives, J.D., B. Messerli, and Spiess. 1997. "Mountains of the world: A global priority." In: B. Messerli and J.D. Ives (eds) *Mountains of the world: A global priority*. New York: Parthenon.

Jodha, N.S. 2000. "Poverty alleviation and sustainable development in mountain areas: Role of highland–lowland links in the context of rapid globalisation." In: M. Banskota, T.S. Papola, and J. Richter (eds) *Growth, poverty alleviation, and sustainable resource management in the mountain areas of South Asia*. Germany: Deutsche Stiftung fur internationale Entwicklung.

Joshi, B.K. 2000. "Development experience in the Himalayan Mountain Region of India." In: M. Banskota, T.S. Papola, and J. Richter (eds) *Growth, poverty alleviation and sustainable resource management in the mountain areas of South Asia*. Germany: Deutsche Stiftung fur internationale Entwicklung.

Kapos, V., J. Rhind, M. Edwards, M.F. Price, and C. Ravilious. 2000. "Developing a map of the world's mountain forests." In: M.F. Price and N. Butt (eds) *Forests in sustainable mountain development: A state-of-knowledge report for 2000*. Wallingford: CAB International.

Kreutzmann, H. 2000. "Improving accessibility for mountain development: Role of transport networks and urban settlements." In: M. Banskota, T.S. Papola, and J. Richter (eds) *Growth, poverty alleviation and sustainable resource management in the mountain areas of South Asia*. Germany: Deutsche Stiftung fur internationale Entwicklung.

Kreutzmann, H. 2001. "Development indicators for mountains." *Mountain Research and Development* Vol. 21, No. 2.

Lhamu, C., J. Rhodes, and D. Rai. 2000. "Integrating economy and environment: The development experience of Bhutan." In: M. Banskota, T.S. Papola, and J. Richter (eds) *Growth, poverty alleviation and sustainable resource management in the mountain areas of South Asia*. Germany: Deutsche Stiftung fur internationale Entwicklung.

Libiszewski, S., and G. Bachler. 1997. "Conflicts in mountain areas – a predicament for sustainable development." In: B. Messerli and J.D. Ives (eds) *Mountains of the world: A global priority*. New York: Parthenon.

Malik, A., and G. Wood. 2003. "Poverty and livelihoods." Unpublished draft paper written for the Aga Khan Rural Support Programme, Pakistan.

Messerli, B., and J.D. Ives (eds). 1997. *Mountains of the world: A global priority*. New York: Parthenon.

Papola, T.S. 2001. *Poverty in mountain areas of HKH region: Some basic issues in measurement, diagnosis, and alleviation*. Kathmandu: ICIMOD.

Prakash, S. 1997. *Poverty and environment linkages in mountains and uplands: Reflections on the "poverty trap."* Thesis. CREED Writing Paper No. 12. New York, USA: Cornell University.

Ruizhen, Y. 2000. "Strategies and experiences in poverty alleviation and sustainable development in the HKH and the Qinghai–Tibetan Plateau Region in China." In: M. Banskota, T.S. Papola, and J. Richter (eds) *Growth, poverty alleviation and sustainable resource management in the mountain areas of South Asia*. Germany: Deutsche Stiftung fur internationale Entwicklung.

Shelley, M.R. 2000. "Socioeconomic status and development of Chittagong Hill Tracts (CHT) of Bangladesh: An overview." In: M. Banskota, T.S. Papola, and J. Richter (eds) *Growth, poverty alleviation and sustainable resource management in the mountain areas of South Asia*. Germany: Deutsche Stiftung fur internationale Entwicklung.

Swiss Agency for Development and Cooperation. 2001. *Mountains and people: An account of mountain development programmes by the SDC*. Berne: SDC.

UNDP. 2001. *Human Development Report 2001*. Oxford: Oxford University Press.

Wood, G., and S. Shakil. 2003. "Collective action: From outside to inside." Unpublished draft paper written for the Aga Khan Rural Support Programme, Pakistan.

World Bank. 2001. *World Development Report 2000/2001: Attacking poverty*. Washington, D.C.: World Bank.

6

Mountain tourism and the conservation of biological and cultural diversity

Wendy Brewer Lama and Nikhat Sattar

Summary

Mountain tourism includes a broad range of recreational, spiritual, and economic activities in diverse mountain regions. It is an economic mainstay of many mountain communities, generating jobs, livelihoods, and tax revenue, and enabling mountain peoples to continue living close to their cultural roots. However, the impacts of tourism on the natural environment and mountain communities can be significant and, in some areas, can threaten biodiversity and the cultural and social amenities that attract tourists.

Mountain peoples are working together at the local level to address immediate concerns but cannot tackle issues such as biodiversity conservation at the ecosystem level. In addition, many mountain societies suffer from the lack of a political voice and power due to marginalization and discrimination. Women and ethnic minorities are often left out of tourism planning and decision-making.

Although mass tourism may further strain already changing mountain cultures, well-planned and managed tourism can give real economic value to the retention of traditional skills, arts, and hospitality, and can generate a variety of tourism-linked livelihoods. The international community has recognized that mountain tourism plays a key role in mountain development. Tourism also holds promise for promoting and contributing to conservation of biological and cultural resources by focusing on sustainability.

Four major principles of sustainable mountain tourism (SMT) are discussed:

1. Tourism should be one, and not the only, means of livelihood and economic development in diversified mountain economies.
2. The benefits and opportunities arising from mountain tourism must flow consistently and in adequate proportions to mountain peoples.
3. The impacts of tourism on biodiversity and cultural diversity must be well documented, minimized, and managed, and a portion of tourism revenue reinvested in conservation and restoration of bioresources, cultural heritage, and sacred sites.
4. Mountain peoples must play an active and responsible role in planning and carrying out mountain tourism, supported by other stakeholders and networks, by government policies and actions, and by technical and capacity-building assistance.

SMT is most successful when it is planned with communities and supported through legislation and policies, capacity building, training and education, and linkages and partnerships with other stakeholders. It is closely linked with a number of other mountain-development and conservation themes, including:

- the need to coordinate infrastructure plans across development sectors;
- promotion of diverse livelihood including tourism to address poverty alleviation;
- strengthening of democratic principles and institutions, and decentralized decision-making through the participatory approach to sustainable tourism management;
- the relationship of tourism to the maintenance of peace and security;
- the role of international and regional cooperation in sustainable tourism;
- opportunities to build linkages between tourism and education, science, and culture;
- concerns over the general lack of legislation, policies, and plans that specifically address mountain issues, and the specific needs of SMT.

The chapter concludes with an elaboration of these principles and a list of detailed actions needed to move forward.

Introduction

For generations, mountain peoples generally survived on an ethic of conservation and tempered use of limited resources. They managed natural hazards, adapted their cultural and social practices, and existed in (what appears to us to be) relative equilibrium with the forces of nature. Many elements of modern economies – including tourism, technology, and access – have significantly changed the choices made by mountain people and have introduced a whole new ethos of mountain development.

Sustainable tourism is a key tool for mountain development and conservation, which can be achieved only through a closely integrated set of development actions that emanate from, and are implemented by, the guardians of mountain resources. Sustainable mountain tourism (SMT) aims high – to serve as a model for environmentally responsible and culturally appropriate tourism, minimizing negative impacts on biological and cultural resources while contributing actively to the conservation and restoration of these valued mountain assets, and profiting local and national economies in general and mountain communities in particular.

The aims of this chapter are as follows:

- to discuss the major issues facing governments, NGOs, communities, and the tourism industry in developing and managing mountain tourism, particularly with regard to threats to biological and cultural diversity[1] in mountain areas;
- to focus attention on mountain communities as the lead agent in planning and managing SMT;
- to develop a framework and next steps for achieving SMT within the context of diversified regional mountain economies, building upon successes and lessons learned made available through the IYM processes.

Recommendations are made as to how the main stakeholders – including governments, NGOs, and community organizations, international and national donor and development agencies, the private tourism sector and its trade associations, universities and research institutes, and tourists and their information networks – can support and assist mountain communities in developing SMT strategies, plans, and processes. Given the balance of capacities and decision-making, governments must take the first steps to empower and facilitate the ability of mountain communities to carry out their roles.

Issues of mountain tourism

Threats to biodiversity conservation

Tourism's impacts on mountain ecosystems and biological resources are of concern on both local and global scales because of the high degree of biodiversity and environmental sensitivity of mountain areas. Unmanaged tourism (including infrastructure and facility development and human activities) can have marked effects on sensitive mountain environments. These include the following:

- *Removal of vegetation* on both a large scale (e.g. for roads, land clearance for ski areas or hotel construction, etc.) and a small scale (e.g. collection of plants, trampling, and disturbance of sensitive vegetation by uncontrolled tourists).
- *Disturbance to wildlife and reduction of wildlife habitat area.* Unless properly managed, wildlife-viewing by tourists can interfere with critical species needs and life cycles.
- *Wildlife poaching and trade in wildlife parts* is sometimes masked by the increased presence of tourists and local guides in wilderness areas.
- *Increased incidence of forest and grassland fires from tourist activities.* With increased numbers of visitors unaccustomed to high fire dangers, forest fires are a real and serious effect of tourism in mountain areas.
- *Degradation of forests from cutting of timber and fuelwood for tourism.* Wood and shrubs are used extensively by tourists and their guides for cooking and heating. In some cases, the collection of wood for fuel is prohibited within protected areas; this often increases rates of harvesting outside the boundaries of these areas.
- *Improper and inadequate garbage and human-waste management.* Tourism generates a high volume of garbage and waste, which mountain communities are unprepared to process. Low temperatures at high altitudes inhibit the natural decomposition of human wastes. Improperly sited toilets pollute mountain streams, affecting water sources downstream as well as the sanctity of sacred lakes and streams.

Some of the environmental impacts of mountain tourism are evident at the local level (e.g. reduction in the forest canopy due to selective tree or limb extraction for fuelwood and construction), whereas other impacts may be evident only when viewed from the bioregional ecosystem perspective (e.g. fragmentation of wildlife habitat and migration corridors due to tourism and other mountain developments).

Threats to cultural diversity

Cultures and traditional ways of mountain life are continuously changing owing to the modernizing effects of education, communications, entertainment, travel, and employment, as well as tourism. Some of the changes include the following:

- the dissolution of distinctive cultural attributes and features, including loss of native languages; disappearance of traditional dress; ignorance of traditional architectural styles and functions; use of legends, beliefs and rituals; support for holy sites;
- loss of traditional cultural values (e.g. honesty, lack of crime, reciproc-

ity, importance of religion, importance of family/community, systems for ensuring equity and well-being among the community);

- changes in gender roles that affect the maintenance of cultural traditions, e.g. cultural or religious practices that require (or are traditionally taken on by) males are now neglected;
- exposure and exploitation of children, creating a culture of begging, which in turn undermines pride and a sense of economic independence;
- a lack of care for sacred mountain sites, important to both highland and lowland cultures, due to the breakdown of traditional community support systems and religious beliefs.

On a positive note, well-managed tourism can give real economic value to the conservation of cultural features, instilling pride in culture and generating a variety of tourism-linked livelihoods. Yet, unless those cultural attributes retained are authentic, tourism can result in the commercialization of culture.

Relationships between biodiversity and cultural conservation

In many mountain regions, traditions of conserving natural resources are closely linked with cultural beliefs and practices. As shown in these examples, understanding and formally recognizing such relationships can be an effective way of strengthening local commitment to biodiversity conservation.

In the Peruvian Andes, "biodiversity and culture are united because the conservation of native seeds and the respect for the diversity of human and animal beings that live around it is part of the (people's) world vision. This vision is broken when occidental visions of "productivity" are imposed, breaking down the natural biodiversity by standardizing cultivars and seeds ... It is important therefore (that) when we think of promoting mountain tourism, we understand the (cultural beliefs) and evaluate the possible impacts. Importance should be given to the *value* of biological and cultural conservation, before considering the economic benefits of tourism."

In Bhutan, as in a number of Himalayan Buddhist cultures, a strong disinclination to the killing of animals, and respect for all life, has been an important factor in protecting wildlife and biodiversity. Throughout the Himalaya, sacred groves of trees have stood for centuries without disturbance, while nearby forests are lopped to bare trunks for firewood and fodder. Bhutan is now reaping the rewards of a well-maintained ecosystem, attracting ecotourists who come to see the pristine forests and wildlife and contribute to local conservation efforts.

Socio-economic and political issues

Mountain peoples often suffer from the lack of a political voice due to limited access and communications, as well as socio-economic marginalization and political and legal discrimination. Many of the issues related to mountain tourism are compounded by the geographic and political isolation of mountain areas and the poor understanding of mountain issues by lowland societies and political leaders. Tired of the toils of mountain life, and seeking better economic and educational opportunities, mountain people move to the cities, compounding overpopulation and poverty in urban areas. Such trends ultimately affect biodiversity, on a bioregional and even global scale.

The topography, location, and political and economic status of mountain areas (including the political power of their political representatives) influence whether they receive government (or donor) attention to infrastructure needs (e.g. Sir Edmund Hillary changed the face of the Everest region for ever by building the Lukla airstrip). Mountain communities without such infrastructure lag far behind in tourism development. However, even when it becomes available, limited access to education and training opportunities means that many mountain people lack sufficient skills and the resources to benefit significantly from tourism. Tourism provides jobs and investment opportunities, but these tend mainly to benefit wealthier households and investors. The trickle-down benefits available to lower socio-economic sectors are generally limited to menial labour jobs; farming and food production; and time-consuming, minimally profitable, handicraft production. Furthermore, a large proportion of the income generated by mountain tourism goes out of the mountains to pay for "imported" materials, food, and services as well as taxes, commissions, and other expenses.

As mountain economies become dependent on tourism, the breakdown of traditional socio-economic systems, skills, and markets reduces the viability and opportunities for diverse livelihoods. Agricultural communities often give up their sustainable practices and the cultivation of a variety of products and shift to growing a small range of crops – often exotic – for sale to tourists. Lack of tourism management (e.g. control of the number of lodges or operators, enforcement of environmental standards) and an oversupply of tourism service providers in a limited market, bring about overcompetition and price wars, leading to declines in service quality and labour practices, and less attention to environmental protection. The impacts of tourism do not affect only those in communities directly involved in tourism: for example, people living in and around mountain protected areas often bear the burdens of tourism, such as increased garbage and security risks, and inflation, but receive little benefit from park entry fees for much-needed local development and conservation.

Linked to tourism is the phenomenon of amenity migration – "the movement of people to a particular region for the vision of life in a quieter, more pristine environment and/or distinct cultural attributes" (Moss

1994). This phenomenon, which is occurring in both developing and industrialized countries, needs to be better understood and addressed in relationship to mountain tourism management. Amenity migrants swell the population of mountain areas, adding to impacts of traffic, congestion, and demands on mountain resources (land, water, construction materials, etc.). The cultural effects of this new class of mountain residents, and their role in conservation and sustainable development, deserve further attention.

Gender implications of mountain tourism

Gender roles and relations often change when tourism enters the local mountain economy. Guiding or transport jobs take men away from the home for long periods of time; some face high risk in mountaineering work, and never come home. The absence of males adds considerably to women's already heavy burdens of household, child-rearing, agricultural, and resource-collection tasks. The additional responsibilities, combined with the relatively low socio-economic status afforded to women and their lack of "economic worth" without earned wages, holds women back even further from pursuing education, careers, and political involvement, and can have adverse impacts on their health, longevity, and (in some ways) their children's welfare.

In some mountain areas and cultures, however, tourism has contributed to higher socio-economic status and independence for women. Their skills in hospitality, cooking, and care-giving to travellers are valuable commodities in tourism. Trekkers in Nepal ranked cleanliness and "friendliness of hostesses" as the priority factors in selecting a lodge. Women also have key roles to play in conservation of natural and cultural resources: village women in Nepal keep the villages and trails free of litter, recognizing the importance of a clean environment to tourism. As these uneducated women gain confidence and economic power, they are becoming more active in community life, taking on leadership roles and raising their status in the communities (Lama 2000).

Framework for sustainable mountain tourism

Tourism has introduced significant economic opportunities to mountain areas and promises to play an active role in mountain development to come. The impacts of tourism on the natural environment and local people can be significant, however, and threaten the very existence of biological and cultural values that attract tourists to a mountain locale. Such concerns evolved into the paradigm of ecotourism, which served as a forerunner to the concept that tourism could serve as a tool for conservation and sustainable development.

With a broader appreciation for the interrelatedness of issues and factors influencing environmentally and socially responsible, marketable, and politically (or institutionally) viable tourism, the concept of sustainable tourism has unfolded and is now a cornerstone of sustainable mountain conservation and development. Sustainability "demands that we adopt a 'systemic' perspective: the perspective that tourism is not an independent system, but a sub-system of larger systems typically composed of interdependent cultural, economic, environmental, political, social, and technological components. One such system is the mountain ecosystem ... In adopting this ecosystemic perspective, we become better equipped to achieve, in strategic analytical terms, the mission of sustaining the integrity of mountain ecosystems, including their human cultures" (Moss et al. 2000).

In seeking to define "sustainability" in the context of mountain tourism, we ask:

1. Does tourism contribute to sustainable mountain development? (and a sub-set of this):
   - How much of a diversified regional mountain economy should tourism constitute?
2. Who benefits, in economic terms, from mountain tourism? (and a sub-set):
   - Are the benefits, and beneficiaries, sufficient to generate support for, and to achieve conservation of, the biodiversity and cultural heritage?
3. Are biophysical resources of mountains degraded as a result of tourism activities?
   - If so, can such degradation be mitigated or reduced to an acceptable level that will sustain natural ecosystems, mountain people's needs, and tourists?
4. Does tourism affect mountain communities and societies positively or negatively (Mountain Agenda 1999)?
   - Are mountain communities sufficiently involved in the planning and management of mountain tourism that they feel a sense of "ownership" and responsibility for its long-term success?
   - What opportunities does tourism bring to mountain peoples who wish to conserve aspects of their traditional cultures and heritage?

From this, the four major components of sustainable mountain tourism can be distilled:

1. tourism as a component of diversified mountain economies;
2. equitable sharing of the economic benefits and opportunities of tourism;

3. conservation of biodiversity and sustainable ecosystems;
4. participation and ownership by mountain people, and support for cultural conservation.

Tourism as a component of diversified mountain economies

Mountain economies that rely solely or largely upon tourism can suffer inordinately if or when tourism declines, as it normally and periodically does, owing to:

- fluctuations in global, regional, or local economies, and people's financial abilities to travel;
- political instability in a mountain area or region, or on an international scale;
- current trends in tourism;
- changes in national policies, regulations, access, or weather that may affect tourists' travel choices or access to or within a mountain area.

When tourist arrivals do decrease, not only do individual tourism entrepreneurs suffer but also the social and political structures of communities that have come to rely on tourism as a primary basis for their economies can be affected. Recent examples include the almost total demise of tourism in the mountains of Pakistan since 9 September 2001 and the significant decline in tourists to Nepal linked to political instability and Maoist activities, where even well-established projects (such as the Langtang Ecotourism Project) have stopped operating effectively. Lodge-owners have agreed to operate only three lodges a day on a rotational basis in each village and to stop the practice of paying commissions to guides. Poorer households, who had seasonal income from tourism, have had to withdraw their children from school as they cannot afford the tuition.

For tourism to be an effective tool for sustainable development in mountains, it must be one (and not the only) means of livelihood and economic development. Such diversification should begin by looking at traditional mountain livelihoods as the basis for potential economic activities, such as high-value agriculture, sustainable forestry, or industries based on non-timber forest products (e.g. herbs, mushrooms, medicinal plants), energy (e.g. hydroelectricity, wind energy), arts, education, etc. As another angle on diversification within the tourism industry itself, development of a domestic tourism industry is called for, drawing upon interests and travel times (e.g. religious, educational, seasonal or holiday travel, cultural exchange) that may vary or complement international travel priorities. Recognizing that amenity migration is a growing phenomenon around the world, cost–benefit analyses should be done, and consideration given to compensating mountain communities for costs incurred (e.g. increased traffic, pollution, demands on resources) while weighing the benefits.

Equitable sharing of benefits and opportunities of tourism

One challenge of sustainable mountain tourism is how to ensure benefits to poorer households who lack capital to invest in, and skills relevant to, tourism-based enterprises. A number of examples of mountain tourism are presented in "Best Practices" (appendix 6.1), illustrating various mechanisms for sharing benefits and opportunities, including a tourism (or bed) tax (wherein tax funds are collected and used for community development needs); rotation of visitors among service providers; and selective training of non-lodge-owning community members as guides. As well, stimulation of a broader, more diversified economic base, with technical and start-up financial assistance, can help to generate livelihood opportunities across socio-economic and gender lines.

Mountain communities are the custodians of the resources and values that make mountain regions so attractive for tourism. If tourism is to be a sustainable means of mountain development, benefits need to flow consistently and in adequate proportions to mountain peoples.

Conservation of biodiversity and sustainable ecosystems

There are many justifications for the conservation of mountain biodiversity, as follows:
1. Mountain regions are "biodiversity hotspots," with high levels of biological diversity at all scales and high concentrations of endemic species; they are vital reservoirs of genetic diversity.
2. Mountain regions function as critical corridors for migrating animals and as sanctuaries for plants and animals whose natural habitat have been squeezed or modified by natural and human activities.
3. The loss of biodiversity has environmental, ethical, health-related, and economic implications.
4. Mountains have a high degree of environmental sensitivity.
5. The declining health of mountain ecosystems not only threatens the survival of highland species and economies but also affects downstream watershed management, water quality and supplies, agriculture, climate, and wildlife-migration patterns.

Although significant threats to biodiversity from tourism are clearly evident in many mountain regions, detailed information about these threats is available from only a relatively limited number of locations, and rarely on a bioregional scale. Such information is essential if local people and other concerned stakeholders are to develop appropriate means for minimizing impacts and, where necessary, to act to restore ecosystems through community-based initiatives supported by scientific and indigenous understanding.

Sustainable mountain tourism implies effective management, which requires identifying, understanding, and measuring the impacts of tourism on biological conditions, and making such information available to decision makers and stakeholders. In areas where the impacts of mountain tourism on biodiversity are less obvious or undocumented, priority must be given to establishing baselines from which to measure change and to developing relevant and realistic methods and means of monitoring impacts. These should be based on an "ecosystemic approach to understanding and management (which) assumes a bioregional perspective, in which the ecosystem is treated as a whole – a symbiotic web of relationships among species and their activities within their spatial territory. It further assumes that we need to plan and act with careful consideration of this interdependent system. It also espouses a holistic intent ..." (Moss et al. 2000). Such an approach "is still rather experimental and problematic, including technical issues and a limited awareness and acceptance by key stakeholders and their institutional processes."

Where mountain ecosystems cross international boundaries, such approaches lead to questions of political jurisdiction. The IUCN has identified 169 transboundary protected areas, between which information sharing and joint management have developed in various ways, from formal, high-level intergovernmental treaties to "bottom-up" field-level cooperation and information sharing between park managers. However, all approaches share common objectives to manage shared natural heritage effectively and to conserve landscape values, ecosystem processes, critical habitats, and a diverse range of plant and animal species (Sandwith et al. 2001).

Monitoring of tourism impacts on mountain biodiversity remains a daunting challenge, burdened by cost and the lack of equipment and trained personnel, time, and accessibility, and the acceptance of standardized easy-to-use biodiversity assessment and monitoring methods. Yet progress is being made and affordable technology is now available (Moss et al. 2000). In addition, much can be done using participatory monitoring methods, not only in terms of measuring change but also in building an understanding among stakeholders of the value of monitoring and its role in adaptive management.

Participatory approach, and support for cultural conservation

The Mountain Forum's electronic conference on "Community-Based Mountain Tourism: Practices for Linking Conservation and Enterprise"

(Godde 1999) revealed the importance of stakeholder involvement and the benefits of a participatory approach to community-based mountain tourism. According to a majority of the conference's 460 participants, successful practices of community-based tourism "appear to be creating a more equitable distribution of tourism opportunities and benefits. All are based on the principles of local control, partnerships, sustainable development, and conservation." Experience continues to show (see appendix 6.1) that participation is a key factor of success, not only in community-based tourism but also in building long-term stakeholder support for sustainable mountain tourism and the conservation ethic it embraces.

One example is the Helvetas Business Promotion Project (BPP) in Kyrgyzstan (D. Raeva, personal communication 2002). The participatory approach is used in initial training and throughout the project cycle, from planning through evaluation. Decisions are made by local people, requiring the agreement of two-thirds of the community group members. All stakeholders develop the yearly plan and participate in the evaluation workshop at the end of the year. Monitoring of plan implementation usually takes place at monthly community group meetings. This approach is derived from a framework called Appreciative Participatory Planning and Action (APPA), which draws from the established methods of Participatory Learning and Action (PLA) and the philosophy and "4-D" cycle of Appreciative Inquiry (AI), developed by Case Western University. APPA was developed by the Mountain Institute and its Asian Program stakeholders in Sikkim, Nepal, and Tibet, shared with NGO partners and associates, and nurtured by each user to address its own needs, from poverty alleviation and conservation to institutional capacity building and women's literacy programmes. It forms the basis of an international training course on "Community-Based Tourism for Conservation and Development," conducted annually since 1998 by the Mountain Institute (Asian Program) and RECOFTC (Regional Community Forestry Training Center, Thailand). Participants from some 30 countries have attended the training course and are now using APPA in their home countries.

The APPA approach identifies and values natural and cultural resources, attributes of mountain areas, human skills, and other mountain-tourism assets as the basis for envisioning, then planning and implementing, a community-based plan for mountain tourism. The planning process and outcome have a strong emphasis on conservation and community self-reliance, building pride and self-confidence, as well as concrete organizational skills to plan, develop, manage, and monitor tourism.

Consistent with the approach of building stakeholder ownership in mountain tourism, sustainable tourism seeks to support the quest of mountain people for:

- attaining legal recognition and respect for their indigenous culture;
- ensuring that mountain tourism contributes positively to the uplifting of their cultural values and heritage; and
- mitigating impacts on cultural and religious tourism sites through proper management.

A framework for community-based tourism for sustainable mountain development

From this analysis, a working definition of sustainable mountain tourism emerges:

Sustainable mountain tourism is that which contributes to meeting current liveli-hood needs, and invests in conservation of biodiversity and mountain cultures, as part of an integrated and participatory approach to sustainable mountain devel-opment that serves the well-being of future generations and maintains healthy mountain ecosystems for the long-term future. Sustainable mountain tourism must be defined for and by each community and culture, in terms of locational attributes and ancestral lineage, as well as within both localized and worldwide perspectives of conservation.

Some of the most promising examples of sustainable mountain tourism have come from the local community[2] level, or at least with the strong involvement of local communities (See appendix 6.1). Reasons for this relative level of success may be that mountain communities:

- are often the best caretakers of their environment, with vast experience and understanding of the mountain landscape and natural systems; as such, they take pride in conveying that knowledge and adeptness to visitors;
- are striving to slow or reverse out-migration of their skilled people and to improve local economies, by developing innovative livelihoods – including tourism – that build upon unique mountain assets;
- are often involved in multiple livelihood activities that vary with sea-son, weather, market demand, and available resources; tourism serves mountain communities best when it is not the only source of economic activity;
- have cultural values and social structures that are complex and based upon a strong sense of reciprocity, which may not respond to forced interventions;
- are best equipped to address the unique challenges to tourism devel-opment and resource management of mountain areas.

Conversely, nationally led tourism planning programmes address the needs of the nation (e.g. to diversify economic development, generate revenue for the national treasury, or subsidize a national transportation system). These objectives may not match the requirements for sustainable mountain development. Furthermore, national governments often lack the commitment of mountain communities to carry out mountain tourism development, giving priority instead to the short-term gains of mass, centrally planned tourism over the long-term benefits of sustainable tourism. Such top-down tourism overlooks the true characteristics or identity (marketed as unique selling points) of mountain tourism, as well as the challenges faced in mountain areas. In contrast, regional tourism (i.e. destination-oriented or based on an ecosystems approach) can play an important role in the planning, management, and marketing of community-based tourism, and is best informed by "bottom-up" tourism planning that realizes and helps strengthen diversity and uniqueness in local tourism products.

The participatory, community-based approach to mountain tourism is, therefore, the recommended path to sustainable mountain tourism. It is by no means a straightforward path, nor free of obstacles: it requires significantly more time (and, therefore, is often more expensive) than a traditional top-down approach and it is highly dependent on having the right staff and partners, with a genuine commitment to, and attitude and skills for, empowering communities. Obviously, communities are not the only stakeholders in community-based mountain tourism: close coordination and cooperation among all stakeholder groups is vital, with clear understanding of each one's role and responsibilities. Figure 6.1 illustrates the roles and relationships among various stakeholders in sustainable mountain tourism.

Best practices for mountain tourism

A collection of "best practices" (appendix 6.1) reflects examples of successful efforts from mountain regions at addressing specific needs of sustainable mountain tourism, including policies, regulations, participatory methods, education and training, investment in conservation, benefit sharing, enterprise development, marketing, codes of conduct, incentives, community empowerment, and partnerships.

No single mountain community or tourism project has put all of these practices together, nor can any serve as a model for sustainable mountain tourism. No doubt, all of these examples face problems and setbacks; none is perfect. But there are lessons in these mini-successes from which other practitioners and policy makers can learn. Because of space limitations, it is not possible to detail the strategies and methods used to understand the success factors and failings of these examples. From this beginning, however – and with greater networking and follow-up exchanges among mountain communities, governments, and NGOs – more in-depth learning and sharing may be possible.

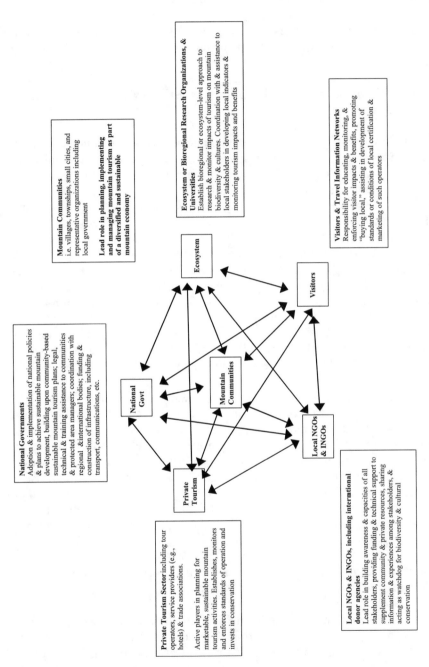

National Governments
Adoption & implementation of national policies & plans to achieve sustainable mountain development, building upon community-based sustainable mountain tourism plans; legal, technical & training assistance to communities & protected area managers; coordination with regional &international bodies; funding & construction of infrastructure, including transport, communications, etc.

Mountain Communities
i.e. villages, townships, small cities, and representative organizations including local government

Lead role in planning, implementing and managing mountain tourism as part of a diversified and sustainable mountain economy

Ecosystem or Bioregional Research Organizations, & Universities
Establish bioregional or ecosystem-level approach to research & monitor impacts of tourism on mountain biodiversity & cultures. Coordination with & assistance to local stakeholders in developing local indicators & monitoring tourism impacts and benefits

Visitors & Travel Information Networks
Responsibility for educating, monitoring, & enforcing visitor impacts & benefits, promoting "buying local," assisting in development of standards or conditions of local certification & marketing of such operators

Private Tourism Sector including tour operators, service providers (e.g., hotels) & trade associations.

Active players in planning for marketable, sustainable mountain tourism activities. Establishes, monitors and enforces standards of operation and invests in conservation

Local NGOs & INGOs, including international donor agencies
Lead role in building awareness & capacities of all stakeholders, providing funding & technical support to supplement community & private resources, sharing information & experiences among stakeholders, & acting as watchdog for biodiversity & cultural conservation

Figure 6.1 Roles and relationships of stakeholders in sustainable mountain tourism

Linkages with themes of other chapters

Among the themes of the other chapters in this volume, tourism is most relevant to the following in terms of the need for coordinated planning and management.

Mountain infrastructure: Access, communications, and energy (Kohler et al.)

Lack of accessibility (including communications) is a defining character-istic of mountain locations. In market terms, however, roads and com-munication networks are the means for linking the tourist to the product. Therein lies the paradox: poor planning for road development can have serious effects on mountain ecology and water regimes (Dasmann and Poore 1979).

Energy – particularly hydroelectric power – is one of the most promis-ing sources of sustainable income available to mountain regions and, if managed properly, can relieve pressure for more damaging resource-extractive activities (such as commercial logging, as in Bhutan). How-ever, like roads, poorly planned energy development can have immense impacts on the natural environment and scenic quality of mountain areas where tourism relies on such. New infrastructure that initially supports tourism can bring about enough negative cultural and environmental changes to make mountain regions no longer desirable to tourists (Godde 1999).

On the other hand, mountains will remain isolated and deprived of basic livelihood needs – including education, health care, political in-volvement, and economic development – unless access, communications, and energy are provided at an appropriate level and form to serve mountain communities' needs. Clearly, careful planning and coordination among mountain communities and government planners and decision makers are needed, with regard to providing needed tourism and other infrastructure as well as ensuring that the impacts of infrastructure do not undermine the scenic and resource qualities that are the basis of a tour-ism industry.

Sustainable livelihoods and poverty alleviation (Parvez and Rasmussen)

Poverty alleviation and sustainable livelihoods are addressed specifically through mountain tourism in terms of "equitable sharing of the economic benefits and opportunities of tourism," but also in striving for "tourism as a component of diversified mountain economies," as well as within "participation and ownership by mountain people, and support for cultural conservation." "Conservation of biodiversity and sustainable eco-systems" is also relevant to poverty alleviation and sustainable-livelihood development in terms of conserving the assets that tourists come to see as the basis for a tourism industry that generates benefits and livelihood opportunities of which all segments of society should partake.

Sustainable mountain tourism, if integrated with a diverse set of live-lihoods, can be a significant contributor to poverty-reduction strategies. With globalization as a given, the mountain community can optimize the opportunities offered by globalization, and national governments can de-velop and implement concrete poverty-reduction steps by attending to tourism development in mountain areas.

To highlight certain needs or target populations, specific activities can be coordinated across sectors, especially in programmes working to-ward diversified mountain-economic development and equitable benefit sharing.

Democratic and decentralized institutions for sustainability in mountains (Pratt)

The participatory approach to building sustainable mountain tourism from the community base upward supports and relies strongly upon democratic principles and institutions and decentralized decision-making. Certainly, cross-fertilization is called for in terms of institutional capacity-building of local organizations, as well as sharing of teaching materials and progress reports.

Conflicts and peace in mountain societies (Starr)

Peace and security, both local and regional, are vital to the development of sustainable mountain tourism, which relies completely upon the move-ment of people unfamiliar with local conditions and on mutual trust be-tween hosts and visitors. It is evident that mountain areas experiencing conflict and a lack of peace and security are shunned by tourists, partic-ularly international tourists, but also by domestic and regional visitors.

Tourism helps promote peace and understanding among peoples and cultures of the world. Tourism exchanges can be developed for these purposes. Secondly, in mountain nations where tourism is a vital part of the economy, national leaders should be made acutely aware of the eco-nomic and other losses incurred due to the lack of peace. In Nepal, Army personnel have been withdrawn from national parks and re-assigned to fight Maoist terrorists, leaving poachers unhampered to slaughter wild-life. In recent months, poachers have killed dozens of endangered one-horned rhinoceros living in protected areas; such losses affect not only the region's biodiversity but also Nepal's tourism prospects.

International and regional agreements and cooperation and sustainable mountain development (Burhenne)

Transboundary tourism and biodiversity conservation require close co-operation among neighbouring nations. Immigration formalities and checkposts must be established and emergency response systems developed for cross-border tourism to occur. Agreements regarding infrastructure development and maintenance, coordinated tourism promotion, responsibilities of service providers, payment methods, etc. must be made before transboundary tourism can be initiated – all of this among neighbours who perhaps speak different languages and have vastly different socio-political or cultural practices. Nevertheless, the prospects for cross-border tourism are great and intriguing to the mountain tourism market.

Likewise, biodiversity conservation and monitoring on a bioregional ecosystem scale require close cooperation and exchange among international and regional bodies, including access to data and legal information. Strong incentives for cooperation and clear understanding through formal agreements need to be in place.

The role of culture, education, and science for sustainable mountain development (Messerli and Bernbaum)

Tourism is not simply a leisure activity for tourists: it gives the opportunity to learn from the people and places encountered, to exchange ideas and perspectives, and to contribute to the protection of places visited for future or others' enjoyment. There is ample opportunity to develop linkages between mountain tourism and the sectors of education, science, and culture in mountain protection and development. For example, close coordination among stakeholders is needed in the development and management of both international and domestic tourism in sacred mountain areas to assure that impacts of tourism are mitigated while benefits accrue to support conservation efforts.[3] Studies abroad, and extension programmes for students and adults, are excellent sources of visitors for mountain communities, as are visiting-scientist programmes. Cultural tourism should be closely linked with, and support, local cultural conservation efforts. Local and international NGOs working across these fields can begin by sharing ideas and plans for specific mountain areas.

Other themes

Regarding linkages between mountain tourism and "legal, economic, and compensation mechanisms in support of sustainable mountain development," and "national policies and institutions for sustainable mountain development," at the Asia High Summit held in Kathmandu in May 2002, participants highlighted the concern that few countries with major mountain regions have policies and strategies that are mountain specific and, hence, the constraints, needs, and opportunities of mountains are not being adequately addressed at the policy and legislative levels. The same is true of many national or regional tourism policies: they are not framed in, or specific to, the principles and needs of sustainable mountain tourism.

With regard to the theme "water, natural resources, hazards, desertification, and the implications of climate change," the concern has been raised in numerous venues (including Kathmandu's High Summit) over inequities in (or the lack of compensatory payment and true economic valuation of) mountain resources, such as hydroelectric energy and medicinal plants, that are exported to benefit down-slope users without due benefit to their mountain guardians. Similarly, a true cost–benefit valuation of tourism resources should be made to underscore the economic importance of tourism and to justify investment in infrastructure development and conservation by governments.

Key principles for sustainable mountain tourism

1. Mountain tourism should be planned as an integral part of sustainable, diversified, mountain economic development that aims to improve livelihood opportunities and the well-being of mountain peoples.
2. Mountain peoples should be given priority and technical or capacity assistance to participate in mountain tourism. Economic opportunities and benefits of tourism should be shared widely and equitably among mountain communities.
3. Mountain tourism development should be governed by laws and regulations, and designed and implemented to ensure the conservation of biodiversity and to minimize impacts on the natural environment.
4. Management decisions should be made on the basis of reliable monitoring of the impacts on biodiversity at the local and bioregional ecosystem levels.
5. Mountain tourism should actively contribute to biodiversity conservation and should build awareness and support for such among visitors and stakeholders.
6. Conservation of the values, traditions, and heritage sites of mountain cultures should be planned and undertaken by the mountain peoples to whom they belong.

7. Sacred sites must be guarded by careful management, by mitigation of visitor impacts, and by educating visitors with regard to proper behaviour and respect for cultural beliefs.
8. Land and resource rights of indigenous peoples should be protected through legal and customary means. Traditional means of nature and biodiversity conservation should be supported.
9. The participatory approach to planning and management should be a principle of sustainable mountain tourism.
10. Tourism should be planned and managed at the community level with active stakeholder involvement.

Action plan for planning and managing sustainable mountain tourism

In table 6.1, (see p. 140) short-term actions are given from top to bottom within each category in a relative time sequence and the order of priority, to implement the principles of sustainable mountain tourism. Long-term actions are paired with short-term actions and do not necessarily run in priority order from top to bottom.

Appendix 6.1: Best practices

Best practices in policy development and implementation

National or provincial-level policies (unless otherwise noted)

- Policy support for *community management of natural or cultural resources and tourism*:
  - Under a "Mountain Areas Conservancy Project" in northern Pakistan, an ecotourism strategy is being developed, using experiences from community management of biodiversity resources in the area. One of the principles to be used is to transfer a fixed percentage of the fees collected to village develop-ment, for use by communities.
- Policy-level *commitment to a participatory process to mountain tourism planning and management*:
  - Kyrgyzstan: Helvetas Swiss Association for International Cooperation was invited by the government to give a training workshop in participatory planning for tourism at the State level, a "chance to introduce participatory planning procedures and ecotourism issues into the tourism policy" (Fueg 2001).
  - Alberta (Canada)'s Provincial Department of Tourism and Multiculturalism provided the guidelines for tourism development according to its provincial

tourism strategy through which communities developed local-area tourism plans. This provincial body encouraged self-regulation and decision-making, as well as broad community participation (Moss 1998 in Mountain Forum/ The Mountain Institute 1999).
- The State Government of Sikkim has started to use the participatory approach in State tourism planning as a result of the demonstrated success of the approach by Sikkim Biodiversity and Conservation at the local level.
• Policy support for an *integrated and diversified approach to mountain conservation and development*, to avoid overdependence upon tourism:
  - Pingwu County Government (Sichuan, China) and Sichuan Provincial Government have supported the Worldwide Fund for Nature (WWF) Integrated Conservation and Development Programme (ICDP) for Panda Conservation, with ecotourism and other enterprise-based livelihoods including improved agriculture, local food and beverage production, handicraft production, and non-timber forest product development.
  - Integrated conservation and sustainable-development strategies developed through consultative processes involving the government and local communities in the two districts of Abbottabad and Chitral in North West Frontier Province and the Northern Areas flanked by the Karakoram/Himalaya/ Hindu Kush ranges includes sustainable tourism for mountain development as a key economic development tool.
• *Coordination among government authorities*, involving policy planning for tourism and related topics such as protected area management and wildlife conservation, trade and industries, transportation, immigration, and finance.
  - Fiji's Koroyanitu National Park Development Programme, centred in the Mount Evans Range (funded by the New Zealand government, and implemented by the Ministry of Forestry and the Native Lands Trust Board) sought to protect cultural heritage and water, soil, and forest resources through the promotion of ecotourism in land-owning villages. Although all operational decisions are at the village level, a larger national framework guides these decisions (Godde 1998 in Mountain Forum/The Mountain Institute 1999).
• *Policy-level cooperation between government and private tourism sector and NGOs* (including trade organizations) in national-level tourism planning and management.
  - Huascarán National Park, Peru, where facilitators from the Mountain Institute brought together national officials, park staff, and hundreds of community and private-sector groups to create a local ecotourism plan. The plan is now seen as "the most comprehensive attempt to manage tourism in the history of natural protected areas in Peru, and the first one specifically tied to a management plan for any unit within the National System of Natural Protected Areas in the country" (Torres 1998 in Mountain Forum/The Mountain Institute 1999).
• National tourism management policies that aim to minimize impacts of tourism through policy standards (e.g. *limiting the numbers of tourists, timing of visits, or group size, or setting operational standards* (with examples of standards or codes of conduct):

- Bhutan government sets a fixed (approx. US$200/day) tourist fare, in effect limiting the number of arriving international tourists by affordability.
- Mustang (Nepal): the government limits tourists to 1,000/year, and charges a royalty of US$70/day to limit numbers of tourists and thereby impacts – but royalties are not reaching the local people.
- In Europe, certification standards and training requirements are strictly enforced for mountain guides, ensuring good safety and professional standards.
- Access to the summits of sacred Himalayan peaks is limited, in respect of local religious beliefs, and is relatively effective.
- Pingwu County policy and now national reserve statutes support Wanglang Nature Reserve's limits on the number of overnight tourists to 50, in order to minimize disturbance to the habitats of the giant panda and other wildlife.

- *Reinvestment of tourism revenues* (e.g. entry fees, lodge or concessionaire royalties, hunting fees) in the conservation of cultural and biological diversity at tourism sites.
  - Park entrance fees: In many mountainous areas, entrance fees are collected as a means of generating revenue for reinvestment in conservation. A significant change in protected-area management policy in the 1980s allowed the Annapurna Conservation Area Project (Nepal) to collect an entrance fee of $13 from visitors, to be channelled into local development and conservation through the King Mahendra Trust for Nature Conservation (Preston/ Mountain Forum 1997).
  - Under Nepal's Buffer Zone Management policy, 30–50 per cent of national park revenues (including tourist entry fees, lodge royalties) are reinvested in development and conservation in communities that lie within the buffer zones and wholly within the national parks. Implementation of legislation is under review.
  - User fees for gorilla watching in Rwanda: Visitors pay fees of $200/day to visit the endangered gorillas in their unique Afro-montane forest homes, thus providing a major source of funding for the preservation of this region and its wildlife. Funds are sent to the National Park office in Kigali and used for patrol and staff salaries, facilities maintenance, and other park needs (Preston/ Mountain Forum 1997).

- Policy *protection of "local" investment opportunities* against domination or profiteering by "outside" investors.
  - Sikkim State policy restricts business licensing to non-Sikkim domicile Indians, including tourism services. TAAS (Trekking Agents Association of Sikkim) bans outside tour operators from joining the association as members, in order to protect its own members' market shares.

- *Policy support for infrastructure development*, including improved access and communications, to remote mountain areas *to diversify tourism destinations* and reduce environmental impacts in heavily used areas.
  - The Government of Nepal has invested in establishing telephone services to every district headquarters in the country and in many trekking villages. Trekkers can call home, and for a rescue helicopter in case of emergency. Tourism entrepreneurs in mountain villages can provide guaranteed available food and fuel supplies, room bookings, etc.

Best practices for practical implementation

- *Participatory learning and planning methods being used*:
  - The Mountain Institute's Himal Program,[4] and local stakeholders together with partner NGOs, developed the Appreciative Participatory Planning and Action (APPA) methodology for community-based tourism planning.
  - Helvetas Swiss Association for International Cooperation has embraced the APPA methodology for tourism planning in Kyrgyzstan, expanding from two initial town project sites into three new sites. Successes include the formation of a community-based tourism (CBT) fund collected as 5 per cent of tourism operators' charges, an almost 50 per cent growth in CBT group members, and improved home-stay standards.
  - Ladakh, India: the Snow Leopard Conservancy has used participatory planning methods (based upon APPA) to plan for home-stay tourism as an alternative livelihood to offset the livestock losses.
  - WWF/Pingwu County ICDP has also adapted APPA for planning ecotourism development in Wanglang Nature Reserve, and in Baima villages. Wanglang staff now use the participatory approach in their own meetings and planning workshops.
  - IUCN – coordinated conservation planning in Pakistan (the Sarhad, Balochistan, and Northern Areas Conservation Strategies), the Himal Project (in Pakistan, India, Nepal, and Bangladesh) and Biodiversity Conservation projects in Nepal and Pakistan have been extremely valuable as practical demonstrations of mountain policy development.
- *Motivating conservation through tourism benefit sharing*:
  - Village home-stay operators in Baima, Sichuan Province (China) donated benches, windows, and materials to the local school.
  - Kyrgyzstan: women in Kochkor and Naryn have formed village tourism committees that operate a booking service and allocate tourists to participating home-stays based upon quality of service/community tourism standards, and visitor feedback.
  - Sirubari Village Resort, Nepal shares benefits among its 100 village households by assigning guests on a rotational basis, while monitoring standards of facilities and service by committee.
  - Villages of Langtang/Helambu (Nepal) allocate 5–10 per cent of lodge and camping charges to pay for conservation strategies such as trail improvement, reforestation, and community toilets.
  - Certain sustainable tourism practices, such as trophy hunting of the ibex and markhor in Pakistan, can add to the economic and attraction value of tourism, but require considerable management effort and strong local participation in planning and benefit sharing.
- *Reinvestment of tourism revenues by non-governmental and private sector* in conservation of cultural and biological diversity in mountain areas, e.g.:
  - Women of Helambu (Nepal) have contributed their own money to operate a cultural museum for tourists; they also perform cultural dances to raise funds for village garbage management and to restore the village monastery (Lama 2000).

- Kangchendzonga Conservation Committee (KCC), in Yuksom, West Sikkim, sells bird lists/guide books, rents binoculars and kerosene stoves, and collects donations to fund environmental education in the community and school.
- Trekking Agents Association of Nepal (TAAN) conducts annual "Eco-Trekking Training" in practical conservation techniques for trekking guides, using trekking-agency membership fees and participation fees to pay for it.
- Mountain "eco-lodges"[5] reinvest in conservation, benefit local people, and employ eco-friendly designs.
- *"Conservation contracts" with the community*:
  - The WWF/ICDP Panda Conservation Project (Sichuan, China) has made "conservation contracts" with Baima villagers to protect the giant panda. In exchange for training and marketing assistance in ecotourism, villagers (some of whom had previously poached panda) volunteer on panda patrols.
  - The Snow Leopard Conservancy (SLC) makes contracts with villagers in Ladakh to protect the snow leopard. Villagers provide labour, stones, and mud to build enclosed livestock pens (rather than killing the attacking snow leopards), while SLC provides off-site materials and follow-up planning for community-based tourism that promotes snow-leopard viewing (www.snowleopardconservancy.org).
- *Education and awareness-building among tourism stakeholders:*
  - The Stevens Village Project (Alaska) helps to educate the community about tourism and alternatives and links the village with information resources and contacts (Mountain Forum/The Mountain Institute 1999).
  - Nepali, Sikkimese, and Tibetan villagers and leaders learned such skills and techniques as composting toilets, and lodge and park management, from each other in "peer to peer" exchanges, thus building relationships across borders.
  - The WWF/ICDP (China) organized a study tour to Nepal for county officials to learn about ecotourism. Repeated awareness-building workshops and meetings have helped to convince leaders to support the development of an ecotourism lodge at Wanglang Nature Reserve and some of the first community-based ecotourism activities in China.
  - Protected – area managers from Nepal and Tibet have come to the United States with the Mountain Institute to learn about tourism and park management in some of the oldest and busiest national parks in the country. Some receive on-the-job training as "Junior Rangers" and go home with new visions of what is possible.
- *Sustainable mountain tourism standards/Codes of Conduct and certification*:
  - Villagers in Ladakh (India) have established criteria for the selection and operation of home-stay operators: these are a minimum of two beds, the serving of simple traditional food, and the maintenance of local culture experiences and ways of life. The majority (83 per cent) of international tourists polled said they thought tourism should benefit local communities (Snow Leopard Conservancy 2001).
  - Wanglang Nature Reserve, Sichuan, China, has a Code of Conduct for how visitors should behave in the panda reserve, in order to reduce their impacts.
  - The Australian National Nature and Ecotour Guide Certification Programme

sets standards for certified guides and offers a certificate for completion of a professional training course.

- NEAP (the National Ecotourism Accreditation Programme) certifies nature and ecotourism sites, primarily in Australia but also internationally, based on very specific criteria for everything from energy use to interpretative skills and effectiveness of tourism-impact management.
- Green Globe 21 is a worldwide certification programme for sustainable travel and tourism for consumers, companies, and communities. The Green Globe standard used for certification is based on Agenda 21. "Green Globe registered companies and destinations will be marketed on line to environmentally conscious consumers around the world."
- The Baima community has set ecotourism home-stay standards (such as clean toilets, and bedroom standards) that not every home can meet; this is the village's way of benefiting non-participating households.

• *Regulation of negative impacts of tourism* combined with practical assistance in implementing policies:
- Government subsidization of kerosene in Sikkim makes it more affordable and available to trekking agencies to reduce the use of fuelwood collection in forests;
- The Makalu–Barun Conservation Project (Nepal) has assisted villagers with loans to establish a kerosene depot to sell kerosene and rent stoves and blankets to porters entering the National Park, to reduce fuelwood use.

• *Skills development and capacity building for sustainable mountain tourism*:
- Nepal has set the standards for trekking services for the region. The Hotel Management and Tourism Training Centre (supported by the government and the International Labour Organization [ILO]) and private companies provide mandatory training for trekking guides. The TAAN and Kathmandu Environmental Education Project (KEEP) have initiated an "Eco-Trekking Workshop" in 1991, which teaches conservation-oriented skills. The training has been taken to Sikkim and Bhutan.
- Several ecotourism and conservation projects in Nepal (e.g. ACAP, Langtang and Makalu-Barun, CCODER) have developed excellent lodge-management training programmes that are given in the village to improve lodge standards and environmental practices. CCODER focuses on home-stay training. Training in energy efficiency includes building low-fuel-using stoves.
- The Mountain Institute and RECOFTC[6] have developed a training course on "Community-based Tourism (CBT) for Conservation and Development." The course uses APPA to promote tourism that is a visitor–host interaction with meaningful participation by both, and that generates economic and conservation benefits for local communities and environments. The international training course has been given for four years (1999–2002), training over 100 international participants from NGOs, government, private sector, and universities from approximately 30 countries. People who have attended the course are using the method in at least seven countries, including Viet Nam, China, Kyrgyzstan, Bhutan, Nepal, India, and Indonesia. A training resource kit in the CBT method has been published and is available com-

mercially, and a trainers' manual is being produced (The Mountain Institute 2000).

- *Successful small-scale enterprises linked with mountain tourism.*
  - Villagers of Langtang/Helambu received small matching grants to establish kerosene depots. The profits from kerosene sales are used for conservation, tourism management, and infrastructure improvements.
  - Local guide services: in Pakistan, village wildlife guides are a group selected, trained, and paid through the Mountain Areas Conservancy Project.
  - Trained naturalist guides in Yuksom, Sikkim are employed by trekking agencies to identify birds and plants and to describe the ecology of Kangchendzonga National Park.
  - Handicraft sales: Nepali village women knit woollen hats, mittens, and socks to sell to trekkers on site. Handicraft retailers and women's development projects in Kathmandu buy handicrafts made by women in rural areas. Transportation and communication, as well as quality control, remain major stumbling blocks to expansion of the production base.
  - Baima women in China have set up a revolving loan programme to enable women to buy yarn to weave traditional belts for sale to tourists. Women could not repay loans because belts were too expensive for the domestic market. WWF/ICDP assisted with the design of new, cheaper products (purses, place-mats, etc.), which, along with home-produced honey, are being sold at the Panda Reserve Headquarters, as well as in village home-stays.
  - Micro-enterprise was successfully used to value both cultural and natural heritage by the Dadia Women's Cooperative in Greece. A women's co-operative was formed in 1994 when the forestry service allowed the women to use the canteen in a recreation area. The village of Dadia then gave them a piece of land to build their own food kitchen. At first, store-owners in the nearby town of Soufli gave them credit for purchasing raw materials, which was repaid once money started flowing in. The women now rent a small building to prepare traditional dishes and sell traditional products. The women were given an opportunity to receive US$114,000 as grant funding but are reluctant to take it because their cooperative is already self-funding and working well (Valaoras 1998 in Mountain Forum/The Mountain Institute 1999).
- *Marketing mechanisms and linkages for small-scale mountain tourism operators*:
  - Effective marketing and promotion of sustainable services and practitioners: the International Ecotourism Society (TIES) is collaborating with its institutional members (tour operators) to promote ecotourism trips during the International Year of Ecotourism. Tour operators commit to a donation to TIES out of trip profits.
  - Other web-based ecotourism organizations (such as the Himalayan Explorers Connections, Adventure Travel Trade Association, Planeta.com, and Ecoclub) offer ecotourism information and marketing exposure for their members, some of which is oriented toward mountain tourism.
  - In Kyrgyzstan, NoviNomad has established market contact with ecotourism operators in Europe and elsewhere to promote community-based ecotourism and nomad tourism in the mountain areas. NoviNomad works closely with Helvetas in developing community-based ecotourism as well.

- CCODER, working with village home-stay operators in Nepal, is a small Kathmandu-based NGO that enables marketing links with local and international tour operators (as well as providing training and project inputs).

Existing and potential partnerships in mountain tourism

- *Partnerships for planning and management*:
  - The Budongo Forest Ecotourism Project in the highlands of Uganda involves the communities of five parishes and is based on wildlife viewing. Partnerships between natural resource managers and their neighbouring communities create a win–win situation in natural-resource management (Langoya 1998 in Mountain Forum/The Mountain Institute 1999).
  - Transboundary tourism epitomizes government-to-government tourism partnerships, and exists between the United States and Canada, Nepal and Tibet (China), China–Central Asian republics, and across mountain borders of European countries. Governments must agree to immigration regulations and enforcement, safety management, and mechanisms for curtailing potential illegal cross-border trade in such items as wildlife parts, medicinal plants, drugs, and weapons. Protected-area and tourism managers, as well as government leaders from Nepal and Tibet, have participated in a number of study tours to border regions to learn from each other and to map out strategies for transboundary resource and tourism management.
  - Partnerships between local communities and NGOs: local NGOs have an important role in working with communities to foster sustainable mountain tourism. Local NGOs, such as Mountain Spirit and the KCC (see above re. Nepal and Sikkim), have functioned both as trainers and planning facilitators, and now (since completion of project funding and activities) provide follow-up assistance to communities in such ways as community development (e.g. the development of a health clinic), environmental education, and monitoring of tourism impacts. NGOs have taken communities "under their wing."
  - Partnerships between local and international NGOs: local NGOs often provide the local expertise (of culture and language), familiarity, mobility, and cost-effectiveness that can serve as the ideal bridge between international NGOs and communities: for example, the East Foundation contracts with the Mountain Institute to carry out fieldwork, training, planning and follow-up, in TMI project sites in the Makalu–Barun area.
  - Waste management on Mount Kenya, Kenya: because of the numerous tourists, problems with litter and human waste are prevalent. Three kinds of initiatives are being undertaken to address the waste problem: (1) informative pamphlets and signs; (2) government-sponsored and private-interest-sponsored group clean-ups; and (3) the dissemination of information regarding this problem by word of mouth (by tour operators to tourists). The key lies in collaboration between interest groups, which currently include the Association of Mount Kenya tour operators, National Park authorities, the Kenya Wildlife Service, National Outdoor Leadership School, the Mountain

Club of Kenya, and UNEP (Carlsson 1998 in Mountain Forum/The Mountain Institute 1999).

- *Partnerships for capacity building and learning*:
  - The Australian Nature Conservation Agency and the indigenous landowners (or Anangu people) jointly manage the Uluru-Kata Tjuta National Park Cultural Centre, Australia. The park houses one of Australia's most popular attractions – Ayers Rock, or Uluru. Over the years, Ayers Rock has become known among tourists as a geological feature to be climbed. To the Anangu people, however, Uluru has tremendous spiritual significance. In an effort to stem visitor climbing, the Anangu and the Australian National Conservation Agency have cooperated in developing the Uluru-Kata Tjuta National Park Cultural Centre. This centre informs tourists of the cultural and spiritual significance of Uluru and the surrounding area (Kelly 1998 in Mountain Forum/ The Mountain Institute 1999).
  - Dig Afognak, Alaska: museums, like visitor centres, can be a vehicle for unifying a community and revitalizing community culture. Dig Afognak was developed to help the Koniaq Alutiiq people recover prehistoric artefacts located on native lands. Now the project is funded by tourists who partake in the archaeological dig and learn about the local culture, geography, and environment. The programme includes lectures for tourists and community members who take part in the dig, combined with valuable hands-on experience (Patterson in Mountain Forum/The Mountain Institute 1999).
  - Partnerships for information sharing/networking: community-based tourism sites and private operators can obtain hard-to-get information about mountain tourism issues and opportunities, as well as market exposure to the international tourism market, by way of websites operated by a number of non-profit ecotourism organizations, including the Adventure Travel Trade Association, Ecoclub, Planeta.com, the International Ecotourism Society, and the Himalayan Explorers Connection/HimalayaNet websites, which provide valuable services to their members and to the consumer.
- *Partnerships in marketing*:
  - Cooperatives are a form of partnership wherein members work together and provide mutual support toward the achievement of a particular goal. The support is often financial. When some members of a cooperative are more successful at selling their product and are earning more revenue, these members have the ability to subsidize other members of the cooperative. Such subsidies work best in communities with an orientation toward communal social organization. Among the Aboriginal people of Australia's central mountain regions, for example, intra-cooperative subsidies are highly effective, owing to a tradition of strong communal bonds. One example is the art centre of Yuendumu, which, like other art centres, is owned by the local community and functions as a cooperative. Entire families work closely together, with the more successful artists subsidizing other artists. Revenue generated from art sales to tourists keeps the enterprise operational. Extra revenue filters down through the rest of the community (Betz 1998 in Mountain Forum/The Mountain Institute 1999).

- Community–private partnerships: Sirubari, a Gurung village in Nepal, has an exclusive partnership with an international marketing agent in Kathmandu. No tourist is allowed to stay in the village who has not come through specified market channels, or the partnership will be dissolved.
- Study-abroad programmes are a fast-growing market. Some study-abroad programmes involve students spending time with mountain families, studying the culture and language and undertaking research for accredited coursework. Participating universities have established partnerships with communities to host students and, in some cases, with international non-governmental organizations (INGOs) to study in their project sites (e.g. the Mountain Institute's School for Mountain Studies).
- See also above with regard to website marketing connections.

Table 6.1 Planning and managing sustainable mountain tourism

Aim	Short-term actions: Initiated during 2003–2005 (may be ongoing)	Long-term actions: Initiated during 2006–2010 (may be ongoing)	Lead responsible stakeholder(s)	Supporting or coordinating stakeholders
Planning and managing sustainable mountain tourism	Develop awareness about sustainable mountain tourism issues among all stakeholders	Ongoing. Conduct research on amenity migration and mitigate impacts	NGOs/INGOs, universities, travel networks, media	Tourism trade associations, governments, communities
	Develop and implement community-based plans for sustainable mountain development, with diversified economies including tourism	Integrate community tourism plans into protected area, biodiversity management, and national economic development plans	Communities and private sector, with assistance from NGO/INGOs and governments	NGOs, protected area managers, development agencies, trade organizations
	Develop a regulatory system with standards for mountain tourism to ensure that sustainability and conservation are addressed	Conduct cost–benefit analysis of mountain tourism. Measure the economic value of conservation to mountain tourism	Government	Tourism service providers/ developers, communities, NGOs
	Develop Codes of Conduct for all users of mountain tourism areas, to minimize impacts and to support local economies	Enforcement of Codes of Conduct by local communities and private sector	NGOs, communities, private sector/ tourism associations, visitor information networks	INGOs, government

140

Capacity building for sustainable mountain tourism	Conduct market research addressing tourists' willingness to pay for conservation, and demand for sustainable services and products	Ongoing. Develop domestic tourism markets	Tourism sector, INGOs	Visitors, NGOs, government
	Develop participatory monitoring and evaluating systems. Conduct training and produce training materials	Continue with an eco-systems approach to monitoring and managing impacts of tourism on biodiversity	Regional research institutes and universities, INGOs	Communities, NGOs
	Coordinate infrastructure-development plans to address tourism needs and impacts	Give priority to disadvantaged mountain areas for use of tourism infrastructure, particularly access and communications	Government	Communities
	Coordinate strategies and plans for sustainable mountain tourism, ecotourism, etc. among INGOs and donor agencies	Establish channels of exchange and conduct conferences/workshops to continue coordination	INGOs, donors	
	Establish and/or strengthen stakeholder user groups to participate in sustainable tourism planning and management	NGOs and community organizations take on roles as trainers	(I)NGOs, government	Communities, private sector

141

Table 6.1 (cont.)

Aim	Short-term actions: Initiated during 2003–2005 (may be ongoing)	Long-term actions: Initiated during 2006–2010 (may be ongoing)	Lead responsible stakeholder(s)	Supporting or coordinating stakeholders
	Empower local communities with legal authority to manage community-based tourism and enforce local conservation policies	Ongoing	Government, communities	NGOs
	Train mountain-tourism operators in eco-friendly practices	Ongoing	NGOs, private tourism sector/tourism trade associations	Government
	Train protected-area and tourism managers in sustainable tourism management, and participatory methods	Ongoing	Protected area managers/gov't, INGOs	NGOs
	Educate decision makers in sustainable tourism and the participatory approach	Ongoing	NGOs	Government

	Support women's and ethnic minority capacity-building needs. Assist women and ethnic minorities to invest in and benefit from mountain tourism through skills training, technical assistance, and capacity building	Conduct exchange-learning and confidence-building exercises with mountain women, including cross-cultural women's tourism	NGOs, governments	Communities, private sector, universities
	Enhance the conceptual and practical capabilities of other stakeholders through education and training	Develop educational materials	(I)NGOs, government	Communities, private sector, universities, etc.
	Initiate partnerships to promote mountain tourism and for information exchange	Strengthen partnerships and cooperation	NGOs, communities, private sector/tourism associations	Government
Biodiversity conservation	Collect baseline data on biodiversity at community level and establish a database at bioregional ecosystem level	Ongoing	Local communities, private-sector tourism operators, NGOs	National governments, research organizations, universities (e.g. graduate students)
	Develop community-based tourism monitoring plans and train communities and local NGOs in monitoring tourism impacts on biodiversity	Monitor tourism impacts on biodiversity, at local and ecosystems levels. Use monitoring results in tourism management and conservation decisions	Local NGOs and communities	National governments, donor/INGOs, private sector

Table 6.1 (cont.)

Aim	Short-term actions: Initiated during 2003–2005 (may be ongoing)	Long-term actions: Initiated during 2006–2010 (may be ongoing)	Lead responsible stakeholder(s)	Supporting or coordinating stakeholders
	Set up systems that contribute a portion of tourism revenues to conservation and restoration of bio-diversity	Set up nature conservation trusts to ensure long-term funds for biodiversity conservation	Government, communities, NGOs	Universities, research institutes
	Attach conservation agreements to community development assistance	Assist communities in taking on long-term conservation responsibilities	(I)NGOs, government	Communities
	Require enforcement of international trade laws on protected species by governments that receive assistance for mountain tourism	Ongoing	INGOs, donors, government	Communities
		Develop and support the availability of low-cost appropriate technology	NGOs, universities/research institutes, government	Communities
	Develop and promote mountain-tourism activities that incorporate environmental education, including student programmes	Develop outdoor learning and nature studies as a form of sustainable tourism	Private sector, NGOs, universities	Communities

144

Goal	Action	Sub-action	Actors	Actors
Cultural conservation and improved well-being of mountain people	Consult with communities on how best to conserve cultural identities, and how mountain tourism can contribute to cultural conservation	Support community-initiated culture-conservation activities. Establish conservation trusts for sustainable funding	NGOs, communities	Private sector
	Strengthen the protection of cultural and religious sites and natural areas, through laws, training, capacity building, investment of tourism revenues in restoration, etc.	Establish sacred/cultural sites conservation plans and funding sources. Gain legal protection status for sacred/cultural sites	(I)NGOs, government	INGOs, donors, government
	Develop mechanisms for sharing economic benefits and economic opportunities equitably	Monitor and document the success of equitable benefit-sharing methods	Communities, NGOs, private sector	Government, financial institutes
	Increase awareness/support for, and adopt policies and laws that ensure respect for, indigenous rights	Share successes with mountain communities	NGOs, communities, universities	Government
	Promote use of locally made products to stimulate local economies and stem leakage	Expand production of locally made products	NGOs, private sector, communities	Government, visitors

Table 6.1 (cont.)

Aim	Short-term actions: Initiated during 2003–2005 (may be ongoing)	Long-term actions: Initiated during 2006–2010 (may be ongoing)	Lead responsible stakeholder(s)	Supporting or coordinating stakeholders
	Enhance mountain peoples' skills and capacities in implementing enterprise activities	Monitor results and adjust interventions as needed	Government and communities together, NGOs, private sector	Universities and research institutes
	Develop build-out projections for amenity migration/tourism growth scenarios; assess impacts and develop plans for mitigation	Ongoing	Government, communities, universities	NGOs, private sector

Notes

1. Cultural diversity: the viable existence of discrete indigenous cultural identities, values and systems (i.e. beliefs, structures, roles, customs, and practices).
2. "Local community" is defined for these purposes to include the population of a hamlet, village, township, or city residing in a definable and discrete geographic area (not necessarily coinciding with a political unit), and its representative government, which functions as the "lowest" level of tourism planning and management in a hierarchy of community to national levels.
3. During the International Year of Ecotourism and of Mountains, numerous publications and guidelines on tourism-issues management were released, including *Sustainable Tourism in Protected Areas: Guidelines for Planning and Management*, which was published in 2002 by the IUCN World Commission on Protected Areas, with support from UNEP and WTO.
4. The Sikkim Biodiversity and Ecotourism Project, Langtang Ecotourism Project, and the Makalu–Barun Conservation Project.
5. Ecolodge criteria ("be designed in harmony with local natural and cultural environments, employ sustainable design principles, minimize use of non-renewable energy resources and materials, benefit local communities through provision of jobs ... and by buying local products and services, benefit local conservation ..., and offer excellent interpretation programs" (Hawkins, Wood, and Bittman 1995).
6. RECOFTC: Regional Community Forestry (Conservation) Training Centre, Thailand.

REFERENCES

Betz, D. 1998. "CBMT: Aboriginal art and community-based tourism." In: *Community-based mountain tourism: Practices for linking conservation with enterprises*. Franklin, West Virginia: Mountain Forum.

Carlsson, U. "CBMT: Keeping Mountain Kenya clean." In: *Community-based mountain tourism: Practices for linking conservation with enterprises*. Franklin, West Virginia: Mountain Forum.

Dasmann, R.F., and D. Poore. 1979. *Ecological guidelines for balanced land use, conservation and development of high mountains*. Gland: IUCN.

Fueg, K. 2001. Personal communication. Kyrgyzstan: Helvetas Business Promotion Project.

Godde, P. 1998. "CBMT: Community-based mountain tourism in Fiji." In: *Community-based mountain tourism: Practices for linking conservation with enterprises*. Franklin, West Virginia: Mountain Forum.

Godde, P. (ed.). 1999. *Community-based mountain tourism: Practices for linking conservation and enterprise*. Franklin: The Mountain Institute.

Hawkins, D., M.E. Wood, and S. Bittman. 1995. "The Ecolodge Source Book for Planners and Developers." In: *The business of ecolodges*. Burlington, Vermont: International Ecotourism Society.

Ives, J.D., B. Messerli, and R.E. Rhoades. 1997. "Agenda for sustainable mountain development." In: B. Messerli and J.D. Ives (eds) *Mountains of the world: A global priority*. Carnforth: Parthenon.

Kelly, J. 1998. "CBMT: Responsible promotion." In: *Community-based mountain tourism: Practices for linking conservation with enterprises*. Franklin, West Virginia: Mountain Forum.

Lama, W.B. 2002. "Community-based tourism for conservation and women's development." In: P. Godde, M.F. Price, and F. Zimmerman (eds) *Tourism and development in mountain regions*. Oxford: CABI.

Langoya, C.D. 1998. "CBMT: Community-based ecotourism development in Budongo." In: *Community-based mountain tourism: Practices for linking conservation with enterprises*. Franklin, West Virginia: Mountain Forum.

Moss, L.A. 1994. "Beyond tourism: The amenity migrants." In: M. Mannermaa et al. (eds) *Chaos in our uncommon futures*. Turku: University of Economics.

Moss, L. 1998. "CBMT: Place in community and the regional perspective." In: *Community-based mountain tourism: Practices for linking conservation with enterprises*. Franklin, West Virginia: Mountain Forum.

Moss, L.A. et al. 2000. "Tourism in bioregional context: Approaching ecosystemic practice in Sumava, Czech Republic." In: P. Godde, M.F. Price, and F. Zimmerman (eds) *Tourism and development in mountain regions*. Oxford: CABI.

Mountain Agenda. 1999. *Mountains of the world: Tourism and sustainable development*. Berne: Mountain Agenda.

Patterson, M. 1998. "CBMT: Our work at Afognak." In: *Community-based mountain tourism: Practices for linking conservation with enterprises*. Franklin, West Virginia: Mountain Forum.

Preston, L. (ed.). 1997. *Investing in mountains: Innovative practices and promising examples for financing conservation and sustainable development*. Franklin, West Virginia: Mountain Forum.

Raeva, D. 2002. Personal communication. Bishkek, Kyrgyzstan: Helvetas Business Promotion Project.

Sandwith, T. et al. 2001. *Transboundary protected areas for peace and cooperation. World Commission on Protected Areas Best Practice Protected Area Guidelines Series No. 7*. Gland: IUCN.

Snow Leopard Conservancy. 2001. *A visitor and market survey for promoting rural and community-based tourism in Ladakh*.

Mountain Institute. *Community-based tourism for conservation and development: A resource kit*. Kathmandu, Nepal: Mountain Institute.

Torres, A.M.E. 1998. "CBMT: Recreational use of Huascaran National Park, Peru." In: *Community-based mountain tourism: Practices for linking conservation with enterprises*. Franklin, West Virginia: Mountain Forum.

Valaoras, G. 1998. "Alternative development and biodiversity conservation: Two case studies from Greece." In: *Community-based mountain tourism: Practices for linking conservation with enterprises*. Franklin, West Virginia: Mountain Forum.

7

Democratic and decentralized institutions for sustainability in mountains

D. Jane Pratt

Summary

Mountain ecosystems present special challenges for the management of large-scale common-pool resources. Apart from the intrinsic challenges of complex environments, particular challenges are presented by the lack of systematic and/or spatially disaggregated information. These not only limit management decisions but, more fundamentally, mean that the problems of mountain people are invisible to governments and development agencies.

Two general approaches to sustainability in mountain regions can be recognized. Traditional mountain cultures practise local sustainability. Such systems are highly place-based, managing for multiple uses of natural resources and self-sufficiency. Linked sustainability systems are more complex, as a primary concern is to ensure the sustainability of environmental services provided to downstream populations while maintaining the rights of upland populations.

Several distinct types of institutional arrangements for sustainability in mountain ecosystems have proved successful. The selection of the most appropriate institutional structure depends upon the following: (1) the degree of local isolation, or the extent of linkages with downstream markets; (2) the intrinsic value of the natural resources and environmental services in a given region. A typology of mountain regions can be developed, using these two axes. For each type of situation, the appropriate development intervention and institutional mechanisms are likely to differ to some extent.

Democratic and decentralized institutions – those that engage the participation of stakeholders, and that recognize and encourage local decision-making – appear to be important for both local and linked sustainability. The challenge, however, is not only in identifying the most appropriate governance structures and processes for a particular area but also in creating incentives to apply them.

Introduction

Moving towards sustainability of natural resources and human economies in mountains depends on meeting a number of challenges simultaneously. These include the understanding of local conditions and the broader impacts of ecosystems and social structures, applying technologies and management regimes appropriate to the area, and arranging incentives that promote sustainable behaviours. A great deal of recent research has shown that successful approaches are more likely when they occur within a framework of institutional arrangements that devolve decision-making power to local levels and involve the participation of interested stakeholders over long periods of time.

Such findings reinforce our growing understanding about the intrinsic nature of mountain sustainability and the special threats posed by inappropriate development. The World Bank has considered many of these issues in its 2003 *World Development Report* on sustainability (Pratt and Shilling 2003).

The objectives of this chapter are as follows:
- to identify the special nature of sustainability problems confronting both mountain communities and those downstream populations that depend on environmental services from mountains;
- to identify the institutional arrangements and management regimes that best support solutions to these problems;
- to identify incentives that might help overcome impediments to more widespread adoption of best practice.

Defining the issues

What makes institutional issues critical in mountains?

On the whole, sustainability in mountains is qualitatively different from sustainability in lowlands. Mountains are characterized by niche ecologies, with a high degree of variability due to altitudinal zonation, and with wide differences in insolation, precipitation, temperature, soils, and other factors within relatively short distances. Managing for sustainability in mountains therefore requires managing for complexity at local as well as ecosystem levels. In addition, mountain communities must deal with an unusually high number of risk factors to ensure adequate livelihoods and sustainability of the resource base on which they depend.

Many traditional mountain cultures have mastered this challenge and remain sustainable as long as they can be insulated from population pressures and external incursions. For example, the average Andean farmer grows over 40 varieties of potato in a single season, along with many other crops, and also has livestock. Farmers manage risk by planting each variety in the appropriate niche, according to slight differences in the plant's needs for temperature, maturation, humidity, and sunlight. One farmer in Peru has been recorded as growing over 100 varieties of potato each year (Zandstra 2001).

How are institutional issues relevant at larger scales (outside the mountains, and between uplands and lowlands)?

The scale of human activity, owing to both population growth and increased per capita production, has increased within the last century to levels sufficient to threaten entire ecosystems. In many countries, the scale and rate of change are threatening severe degradation and/or destruction of key functions of mountain ecosystems. The threats stem primarily from three sources: (1) the mountain communities themselves, which in some cases are unable to continue sustainable management practices owing to population pressures and incursions into their territory; (2) external pressures from downstream populations and/or industrial interventions to extract resources that are causing massive disruptions in mountains; and (3) global factors, such as climate change, that are altering the ecosystems themselves. We can think of the first type of threat as a threat to *local sustainability* and the second type of threat as a threat to *linked sustainability*, where the disruption to sustainability comes from expansion of inequitable linkages between upstream and downstream communities. The third type of threat is due to cumulative failures of sustainability, which can be addressed only through international institutions and coordinating mechanisms.

How does the importance/relevance of the issue vary from one region to another?

Increasing evidence suggests that participation and decentralized power to make decisions about issues affecting communities is critical for both

(1) local sustainability *in* mountains for the benefit of mountain communities themselves, and (2) linked sustainability *of* mountain and lowland ecosystems, to ensure the sustainability of environmental services for downstream users. Framed in a way which is useful in considering policy recommendations, incentives are needed to promote collective action for sustainable management of common-pool resources that exist within mountain communities and that span upland and lowland communities (common-pool resources are defined as those that generate finite quantities of outputs that may be used by a number of people, but where use by one person diminishes the availability of the output to others). Generally, these resources produce a variety of outputs and have higher total productivity if managed as a whole than in small units. Such common-pool resource systems are prevalent in mountains, complementing privately owned resources such as agricultural land. Because mountain ecosystems are inherently fragile, overuse or failure to manage common-pool resources sustainably can result in rapid degradation (or even collapse) of the resource base.

This suggests that institutional arrangements appropriate to sustainable-resource management and development are likely to depend on two sets of factors: (1) the natural-resource endowment of the region itself, where natural resources are the basis for traditional subsistence economies; (2) the extent of linkages between upland and lowland communities. We may think of a simple categorization such that a mountain area can be "high" or "low" with respect to the environmental services it provides (watershed protection, ecological significance, recreational value, air quality, stabilization of weather patterns), and "high" or "low" in its degree of connection with lowland populations and markets. Note that "environmental services" here are distinct from "natural resources." This permits us to consider where natural resources are important to maintain *in situ*, because of the services they provide (biodiversity, watershed protection), and where they are important because of their export value for downstream and/or global markets (as is the case with minerals). We can examine, then, where environmental services and market economy values are complementary and where they are in conflict. The types of development interventions and institutional mechanisms appropriate to each situation are likely to differ somewhat, as follows:

- *Low environmental services value/low linkages* to downstream populations and markets: mountain areas of this type are not common; however, where they do exist, they are likely to be those with the greatest economic poverty. They may also be characterized by unusual cultural richness due, in part, to their relative isolation. Such is the case in some portions of Tibet Autonomous Region of China and the highlands of

Ethiopia. Appropriate institutional arrangements at national and regional levels may need to focus on subsidies or other welfare payments to ensure that these communities have access to basic social services and can preserve their cultural heritage. In these circumstances, a major problem is how support for improved productivity and basic subsistence can be designed to work through and strengthen local decision-making and governance.

- *Low environmental services value/high linkages*: areas where ecological and biodiversity values are low, but linkages are high where other resources, especially minerals, are plentiful, as with the arid mountain regions of the Andes. Here, conventional policies such as environmental- and social-impact assessment, avoidance, and mitigation should suffice; nevertheless, too often, they fail. How can institutional arrangements and policies provide incentives for action?
- *High environmental services value/low linkages*: where biodiversity and endemism levels are high, there is often high potential for sustainable livelihoods based on natural-resource management, traditional knowledge and skills, and new opportunities for ecotourism that benefits local communities. Nepal and Costa Rica have developed extensive ecotourism markets. How can institutional arrangements in such circumstances promote the partnerships that are needed between national governments and local people? How can countries create and manage parks and protected areas in the interest of local people who must be involved in and benefit from conservation? What is needed to expand best practice? And how can fluctuations in tourist visitation be managed?
- *High environmental services value/high linkages*: situations where both high environmental value and a high degree of linkage to downstream markets exist present the most contentious challenges for sustainability in mountains. They involve difficult trade-offs between conservation and development, and there are few "win–win" solutions. The damming of the Biobío River in Chile to provide clean energy to coastal populations clearly sacrificed one environmental "good" (biodiversity and indigenous culture) in exchange for an economic one (cheap energy) and an environmental one (clean air) (Brown 1998). How can mountain communities gain access and voice in the political decision-making when their relative power is so unequal?

What are the gender implications?

Traditional mountain cultures may well discriminate against women; however, at the same time, there is less gender-differentiation of labour in many of these communities. Since family members must be able to replace the labour of any absent member, most individuals can (and do) perform multiple roles without a high degree of gender specialization. Increased linkages with lowland markets create particular threats and opportunities for women and girls. On the one hand, their political access and voice is typically further eroded in interactions with institutions from downstream political centres and markets; at the extreme, they are vulnerable to being deceived into prostitution and even slavery. On the other hand, increased exposure to the norms and practices of urban women, and to visiting tourists from the lowlands, can increase opportunities for education and improved livelihoods for mountain women. What is clear is that, unless institutional arrangements are designed deliberately to address gender concerns, mountain women are likely to be left further behind (Byers and Sainju 1994).

What do we know?

Academic comparative research has generated a great increase in our knowledge about development effectiveness and sustainability in recent years. A major gap remains, however, between knowledge and practice, especially in mountain regions, which often are the last to benefit from replication of promising approaches.

What don't we know – and why? How does our knowledge vary from one region to another?

As with most information regarding mountains, there is a great deal of new research and information about particular cases. The most important gap is that such information is neither systematic nor disaggregated spatially. We know, for example, that mountain people tend in many ranges to be poorer and less well nourished than their lowland countrymen. We know why this is so in most cases: agricultural productivity is lower and caloric requirements increase with higher altitudes. What we do not have are spatially disaggregated data that indicate how relative well-being changes within a given country or region. This lack is critical, because it makes the problems of mountain people invisible to governments and development agencies: they are hidden in the general statistics on poverty, and these are not place specific. Equally important, government agencies and donors are challenged in making the best use of scarce resources. Ending poverty everywhere, for all people, is an overwhelming task. With spatially disaggregated data, interventions can be targeted to those who need it most, and can be designed to take account of the specific circumstances faced by local communities.

Practical examples

Analyses identifying the particular character of institutions required for sustainability in mountains are found in long-term studies of institutions for forest-resource management carried out by the International Forestry Resource and Institutions Research Program of Indiana University. Poteete and Ostrom (2002) argue that institutions for resource management are products of collective action. The conditions favouring institutional success depend, they find, on two sets of factors – attributes of the resource and attributes of the users. Many of the user attributes they cite are relevant to mountains and seem to be conducive, particularly or uniquely, to management by democratic and decentralized institutions. These include salience of the resource to the users; common understanding of the resource; trust and relationships of reciprocity; autonomy from external authority; and prior participation in local associations with each other. All of these are attributes of users that argue in favour of subsidiarity in decision-making, as well as institutional arrangements that are participatory and representative.

Local sustainability

Traditional mountain cultures – those demonstrating local sustainability – are highly place-based and must manage for multiple uses of natural resources. Because of their remoteness from markets, in many cases, they also must manage for self-sufficiency within a given locality. Such cultures tend to manage intensively, rather than extensively; to make use of a large and diverse number of species; and to control for a large number of risk factors. In doing so, their management practices begin to approximate to the serving of the sustainability requirements of the ecosystem in which they live.

An important characteristic of local sustainability thus involves inherent incentives: such communities have little (if any) alternative to depending on the resources that surround them. Elements of successful institutional approaches in such cases include providing for local control over key aspects of resource use and allocation; participation and/or representation of stakeholders to ensure that communal resources are managed for equitable use and benefit of all members; and mechanisms to manage risk. When these elements are present, communities can achieve local sustainability – provided that they are insulated from external encroachment.

Local communities' livelihoods, and the sustainable management of the resource base on which they depend, thus require institutions that are adapted to local circumstances and that build on local knowledge and needs. Such requirements are intrinsically inimical to large-scale, centralized management from a distance. In the example from Nepal, below, initially centralized property ownership was coupled with ineffective capacity for administration in the remote mountain areas under state ownership, leading to degradation. Giving effective control and use to local communities resulted in shared agreement on regulations, which led to restored sustainability of the forest resources and improved livelihoods for poor communities.

Nepal: Policy incentives for local decision-making

Following independence, the Government of Nepal assumed national control of forests that had been managed previously under a system of community control. Over the period when resources were owned and managed by the government, deforestation increased at an alarming rate. As government was remote from the resource, its ability to exercise effective control was weak: villagers exploited forest resources illegally because penalties for doing so were unenforceable. In 1978, the government created an innovative programme of community forestry, under which large tracts of national forest were turned over to forest-user groups that could produce a sustainable management plan. Although government retains formal ownership, conservation and sustainable use are determined by decentralized, local institutions that have authority to set and collect fees for use, and to impose fines and penalties for violations of established rules for sustainable harvesting. As a result, not only has forest health been restored in areas managed by the user groups but also communities are now generating money from the forest resource through sales (and fines), while also meeting their own needs for timber and non-timber forest products. Challenges remain, however, because active management of community forests has led in some cases to increased populations of predators, hiding places for Maoist guerrillas, conflicts among users, and even increased pressure on more distant government-owned forests that are not as well protected. There is also a risk that community control empowers élites, leading to further marginalization of the poorest. These problems may abate when community forests produce enough to contribute more to local livelihoods and when villagers receive additional training and assistance in forest management, such as appropriate mapping; nevertheless, countervailing population pressures may offset these gains (Preston 1997; Baral 2002; Bhandary 2002; Stevens 2002; Timsina 2002; Upadhaya n.d.).

Enabling local sustainability requires collective arrangements that foster self-sufficiency within a given region, based on complexity, multiple use, and a large degree of diversity in production and management of resources. For local sustainability, such institutions must be based, above all, on ensuring rights to common-pool resources, and on engaging the ongoing participation of all stakeholders.

Often, mountain communities, left to themselves, have devised such common-property systems on a voluntary basis over time and have thoroughly integrated these practices into their traditional cultures. Such is the case with harvesting of non-timber forest products among the Maori people of New Zealand, for example, where access to resources in mountainous areas is strictly controlled by social and religious sanctions. Similarly, in the Andean tradition of mutual reciprocity (*ayni*), relationships between communities and natural resources are reinforced by sacred traditions that regulate use of communal resources in mountains (The Mountain Institute [TMI] 1998). Reviewing the history of conflicts in the Altiplano region of the Andes, for example, Hernandez (2002) concludes that "the most important sources of conflict [were] the installation of systems of holding and owning land that generated extreme social inequalities; [and that public policies] persisted in maintaining these inequalities and exacerbating conflicts." In other words, the disruption/ displacement of traditional systems that resulted from colonial rule led to centuries-long conflicts; and these conflicts have abated in proportion to increasing recognition of traditional rights and respect for traditional culture. In a meeting of community associations in 1996, peasant organizations agreed on a declaration of principle that land titles should be granted to communities rather than to individuals, land use would be governed by communal assemblies, and that family possessions would be distributed on the basis of rules of usufruct of communal lands. Such structural changes have resulted, according to Hernandez, in substantial improvements – greater social and economic equity, emergence of communal associations and institutions, and a sense of dignity and respect for the culture of traditional Aymara and Quechua communities.

In the Andean and other cases, where colonial rulers have disrupted traditional practices, external intervention may be needed to restore or reintroduce effective community controls, as in the example of the revision to Nepal's forest policy, referred to above.

Land-management regimes are not the only basis for sustainability, however: integral to such systems in many cases were the technologies that accompanied them. The *waru waru* system of the Andean Altiplano, for example, combined a system of communal labour management and ownership with an ingenious technology of water-harvesting and terracing that was ideally suited to the shallow soils and water scarcity characteristic of the region. Re-creation of this communally based, decentralized, and participatory system has generated encouraging results – improvement in crop yields, family food security, and general living conditions, employment generation, and (perhaps most telling) reduction of out-migration (Benavides et al. 2002).

To prevent us falling into the trap of romanticizing traditional cultures, however, Lopez (2002) reminds us that traditional cultures can also be dramatically unsustainable. In Colombia, a culture that identified its prosperity by the number of descendants resulted in local environmental degradation that led to both internecine conflict and out-migration, resulting in substantial population pressures in areas of new settlement as well as in the areas being vacated. When government responded by trying to create protected areas to prevent further degradation, resettlement of peasants created further conflict, as well as increasing the impoverishment of relocated families. In such cases, governments tend to react by focusing on specific causes rather than by analysing the problem holistically; this then leads to the proposal of solutions that are narrowly technical and fail to take account of economic, social, or political dimensions. Inevitably, Lopez argues, violence then emerges because underlying factors of inequality and poverty are not addressed.

Linked sustainability

In cases of linked sustainability, the primary concern is to ensure the sustainability of environmental services provided to downstream populations. Here, reciprocal rights and obligations must be recognized. Populations living in the foothills of mountains have rights to clean water, protection from disaster, recreation, and other amenities, for example. Upland populations have rights to social services and decent quality of life, including respect for their culture and traditions. Three examples are given below.

New York City: Communications and market mechanisms

Failure to control the quality of runoff from dairy and other farms in upland watersheds resulted in a serious threat to the quality and quantity of water available to over ten million downstream users in New York City, USA. Residents were faced with the looming need for massive new investments in water treatment, estimated to cost over US$6 billion. Instead, the city invested US$35.2 million as an incentive for some 400 upland farmers to install pollution- and/or erosion-abatement measures. Because success (and payment) depends on participation of at least 85 per cent of the upland farmers, the programme was designed to be voluntary and to be run entirely by the farmers themselves. They met as a decentralized, 21-member, democratically elected Watershed Agricultural Council to decide on priorities for allocation of the city funds: as a result, water quality and quantity have been ensured, and the massive investment in new water-treatment facilities has been avoided. Equally important, the sustainability of farming systems in the upland watershed has been enhanced, and long-standing distrust between upstate farmers and the City of New York is being replaced by new bonds of trust and understanding, leading to more equitable political decision-making in other areas (Preston 1997).

Guinea: Protecting forests through co-management

The National Directorate of Waters and Forests (Direction Nationale des Eaux et Forêts; DNEF) is legally responsible for management of Guinea's 113 national classified forests. Although most of these forests were classified by the French colonial regime in the 1940s and 1950s, owing to limited government resources they have received little active management. Many have become degraded as a result of years of uncontrolled animal grazing, wildfire, clandestine timber cutting, and illegal encroachment. New management approaches are needed to stabilize and improve the condition of these forests to ensure that they meet objectives of protection of watersheds for both Guinea and neighbouring countries, biological diversity, and provision of needed forest resources.

Since 1992, the United States Agency for International Development (USAID) has been working with DNEF to improve natural-resource management in the Fouta Djallon highlands of Guinea through co-management, aimed at sharing management responsibilities as well as benefits between the national government and the local population. In 1999, DNEF signed the first five-year contract with an inter-village committee to co-manage the Nialama Forest, which is approximately 10,000 hectares in size and is surrounded by about 30 villages and hamlets, home to more than 5,700 people.

The Nialama forest-management plan calls for local involvement in forest protection – in fire management, protection of wildlife and wildlife habitat, and protection of sensitive ecological areas. In exchange, the local population will be allowed limited use of forest resources, to develop an agroforestry system; to cultivate lowlands; and to implement sustainable commercial harvesting of timber, firewood, bamboo, and other non-timber products.

The challenge is to implement the plan. To do so, all involved stakeholders will require training, technical assistance, and other support both for technical forestry-management issues and for organizational management. Strategies are currently being developed to strengthen the institutional, organizational, and financial management capacity and to help determine viable cost-recovery systems for agricultural products.

Although implementation has hardly begun, the Guinean Government and USAID feel that this approach is quite promising. Therefore, they have agreed to replicate this approach elsewhere in Guinea, aiming to have 100,000 hectares of classified forest under co-management by 2005. In 2000, this new approach was successfully applied to the Souti Yanfou and Bakoun classified forests, covering an additional 40,000 hectares, and compressing a planned five-year start-up period into the space of one year (Latigue 2002).

Costa Rica: Hydroelectric investment in upstream stewardship practices

In Costa Rica, private landowners are compensated by the National Government and Energia Global, a private hydroelectric company, when forest cover is maintained or increased in watershed areas. To pay for these services, the Government of Costa Rica established a fund, consisting largely of a 5 per cent tax on fossil fuel, through the National Forest Office and National Fund for Forest Financing (see Koch-Weser and Kahlenborn, this volume, ch. 4, for more detail). Payments are based not on the value of the hydroelectric services but on the opportunity cost of forgone land development, which is primarily cattle ranching.

In a number of cases, this initiative has led to improved forest conservation and decreased impacts on sedimentation and stream regularity. However, in Arenal, even the combination of government payments for reforestation and the elimination of ranching subsidies did not provide enough incentive for farmers to reforest steep slopes. In addition to generating greater upstream-landholder returns, the increased water-yield resulting from the deforestation outweighed the downstream costs of sedimentation, because the yield was of direct benefit to a hydroelectric facility (Chomitz, Brenes, and Constantino 1998).

In the cases of New York City, Guinea, and Costa Rica (see above and Koch-Weser and Kahlenborn, ch. 4, this volume), the actions of upstream users prior to the interventions described were suboptimal from the standpoints of both local and linked sustainability. The management practices of individual farmers were slowly degrading the natural-resource base, so that production levels were decreasing; at the same time, environmental degradation was causing harm at lower elevations. Downstream communities depending on maintenance of environmental services from their neighbouring mountain ecosystems risked being seriously disadvantaged. In both cases, innovative individuals and institutions were able to identify jointly beneficial mechanisms to compensate upstream users for better management of the natural resources, so that downstream degradation was reduced or avoided. This involved getting both upstream and downstream communities to agree to better management of a joint common-pool resource in the entire watershed, and to share the costs and benefits. Participation and decentralized decision-making, thus, were spread across a larger common resource.

Again, there are many opportunities for replicating positive examples of linked sustainability, where downstream communities have invented mechanisms to compensate upstream-resource management for improved management – to the benefit of both upstream and downstream parties and to the benefit of the sustainability of the ecosystem and its environmental services.

For linked sustainability, the successful institutional arrangements described above incorporate improved communications and information-sharing among the various parties involved. In addition, institutions to address these challenges have provided market arrangements and service payments. As with the New York City example, successful arrangements must incorporate a willingness to pay on the part of beneficiaries. Thus they, too, reflect key elements of decentralized and democratic institutions.

In the Arenal area of Costa Rica, failure to connect upstream and downstream users resulted in serious degradation of the watershed on which all depended. In Guinea, a partnership between government, donor, and local communities provides a promising model for sustaining mutual benefits through contracting mutual obligations. Using decentralized, democratic institutions to open communications and information channels, and thereby to create a market linkage between the two, has created an excellent example of linked sustainability.

Perhaps the best – and, too often, overlooked – examples of linked sustainability, however, are those that involve the use of government's conventional regulatory and fiscal authorities to protect common-property environmental resources. One of many good examples is found in the constructive partnerships with local communities and universities developed by the US Environmental Protection Agency (EPA) to protect the sensitive mountain environments of the San Juan range in the southern Rocky Mountains (Reetz 2002). Initial efforts to protect wetlands from illegal fill by developers helped create trust with local communities. Subsequently, EPA was able to build on this trust to insist on "least damaging" alternatives to proposed expansion of ski resorts. EPA funding of research by the nearby University of Colorado led the county to adopt additional land-use codes to protect areas threatened by development; in addition, provision of incentives to local communities has resulted in the adoption of further local controls to protect areas that are critical sources of water.

Part of the success of the previous example is the result of government using its convening power to bring stakeholders to the table. Peru has developed this capability to the level of a national institution: round tables for consensus building provide decentralized, highly participatory community groups with opportunities to set their own priorities, action plans, and budgets, which are negotiated through a General Assembly of the community round tables and enforced through a representative Council. According to Tupayachi (2002), "This has spurred development by mobilizing resources for collective action ... In the process, it has strengthened traditional cultural values and practices ... and solidarity, both within and between communities."

Whereas the Alpine Convention provides an extraordinary example of efforts to promote sustainability through an international legal instrument (see Burhenne, ch. 10, this volume), Turkey may be the only country to have incorporated specific provisions in its Constitution to encourage collaboration between government and rural communities designed to protect forested watersheds (Duzgun 2002). In the 1970s, the Ministry of Forestry embarked on a programme to decrease pressure on forests by supporting development efforts in partnership with local communities. The framework developed was enshrined in the Turkish Constitution in

1982, which provides that cooperation between the State and forest villages should be maintained, and furthered through application of related laws. This has led to a wide range of programmes to provide income generation through subsidized credit, comprehensive planning, and multi-purpose projects. Despite continuing challenges, the result has been significant gains in both environmental protection and improved livelihoods for communities.

In Lesotho, a very different case exists: commissioning of its major hydropower station in 1988 created the ironic situation of making the country self-sufficient in power generation while the poorest communities, living in the remote headwaters regions, remained unconnected to the grid (Mphale 2002). In this case, the revenues generated from selling power to neighbouring South Africa created the opportunity for government to support upstream communities in shifting from reliance on increasingly scarce biomass fuels to petroleum products or, preferably, renewable sources such as wind energy. Failure to do so, as we have seen in other cases, would be likely to lead to increasing inequality and poverty in mountain communities, spawning further environmental degradation in upland watersheds and the emergence of political protests.

Finally, in linked sustainability, the importance of integrating local culture and social organization is underscored by a government and donor project in the Atlas Mountains of Morocco (Crawford 2002). The project aims to halt environmental degradation and to increase incomes of local communities by encouraging tourism and protected-area management. Not only does the project have laudable goals, it also appears to have worked to involve local communities in project implementation. However, the traditional culture, while clinging to the notion of patrilineal organization, in fact organizes communal labour into more numerically balanced units that constitute the fundamental social institutions of the village. These structures have their own inherent elements of fairness and discrimination. The laudable sustainable-development goals of the project, in this case, are unlikely to be realized unless these complex social factors, in addition to the environmental and economic ones, are understood and addressed.

Implications: Best practice

Implications for policy development and implementation

In the case of local sustainability, with low degrees of dependence on outside linkages, successful institutional approaches can follow a relatively simple structure: ensuring participation; promoting community

property-management regimes; supporting traditional culture; and protecting effective systems from external encroachment that disrupts communal ownership rules and management practices. Improved livelihoods
for mountain people in such systems will depend on technical assistance
and, in some cases, transference of payments to ensure that basic social
services are provided. For local sustainability, such institutions should
focus on promoting common-pool property regimes and enhancing participation of stakeholders.

Linked systems are inherently more complex, given their high degree of
interdependency. Institutional structures, in such cases, must provide for
negotiating, monitoring, and enforcement of agreements among stakeholders, who are provided with countervailing power to ensure that upstream and downstream interests are balanced. In the case of linked sustainability, institutions must be designed to address communications and
information needs and to develop markets that can ensure appropriate
and equitable payments for environmental services. Because of their increased complexity, and because of the disproportionate power relations
often involved (e.g. with extractive industries such as mining, water supply, and timber), institutional arrangements for linked sustainability are
more likely to be formal, relying on rule of law and markets, and on
statutory bodies that operate within the framework of national policies.

Implications for practical implementation

Because mountain ecosystems generally cover large areas, and because
traditional mountain communities are typically small, many individuals
and families can draw upon common-pool resources simultaneously without placing undue pressure on them. In these cases, local sustainability
can be managed through creating simple institutional frameworks that
provide for local common-property rights, local autonomy in decision-
making, and shared responsibility among members of the community.
When multiple communities utilize common-pool resources, or when in-
terdependencies are recognized and addressed (as happens with linked
sustainability), common-pool resources appear to be successfully managed when they incorporate key features such as decentralized decision-
making, recognition of the common-pool nature of the resource, participation, and shared responsibility among stakeholders. Problems that
arise – as, for example, when extractive industries or other incursions in-
trude on traditional systems – often appear to result from failure to recognize and provide for shared responsibility for common-pool resource
management. Local sustainability is irrevocably disrupted, but is not replaced by management regimes that would encourage a new, linked, sustainability to emerge.

Returning to our earlier typology, we can tentatively identify the kinds of institutional arrangements that are likely to be most practical in the different cases, and the issues that are likely to create challenges in implementation, as follows:

- *Low environmental services value/low linkages.* Appropriate institutional arrangements at national and regional levels may need to focus on subsidies or other welfare payments to ensure that these communities have access to basic social services. Improvements in local production are also crucial to avoid dependency. Research and extension programmes, such as those of the Consultative Group on International Agricultural Research (CGIAR)'s International Potato Center, are most promising because they target high-altitude production systems. Improved food security then provides the base for greater opportunities for human-resource development and for strengthening local decision-making and governance. At the same time, improved food security and recognition of the need for special support can provide hope for a better future, reducing the rate of out-migration of youth. Because of their remoteness and isolation, decentralized institutions are essential to sustainability of any interventions in such areas. Communal control of resources, and active participation in research and extension services are needed.

- *Low environmental services value/high linkages.* Here, conventional policies such as environmental- and social-impact assessment, avoidance, and mitigation should suffice; nevertheless, too often, they fail. As we have seen in examples above, institutional arrangements and policies must first be put in place, and oversight of individual extractive industry projects must be designed, both to provide incentives for compliance and to ensure enforcement of regulations. Promising results have been obtained using participatory monitoring and evaluation approaches, and adequate involvement of local representation in decision-making. This is particularly important during crises, as when accidents occur during the development of mining projects. Participatory monitoring and evaluation should also continue to be integral to such projects.

- *High environmental services value/low linkages.* Highly developed institutional mechanisms have been tried and proven in many countries and regions, including national parks, protected areas, international treaties and conventions (such as Man and the Biosphere, and World Heritage programmes of UNESCO, Alpine Convention). In such circumstances, formalized involvement of local stakeholders can help promote the partnerships that are needed between national governments that create and manage parks and protected areas, and local people who must be involved in and benefit from conservation. Repatriation of fees to local communities to support projects that respond to community priorities has helped link conservation and sustainable enterprise development.

- *High environmental services value/high linkages.* These involve difficult trade-offs between conservation and development, and there are few "win–win" solutions. Negotiations are needed, and incentives are needed to ensure that mountain communities and interests are adequately represented. To gain access and voice in the political decision-making when their relative power is so unequal, mountain communities need support to be able to build their own constituencies and to equalize power through legal protections to their land and resource rights.

Existing and potential partnerships

A key element of successful institutional mechanisms is to incorporate incentives that foster stakeholder participation. In each successful example, stakeholders, or their representatives, were involved in designing and implementing solutions over an extended period of time. This was possible because incentives were created to make it in the best interest of stakeholders to continue working together. Often, such incentive structures appear to be maintained as much by social incentives and "soft sanctions" that depend on peer pressure for compliance as they do on formal sanctions. Interestingly, the successful cases all make explicit provision for representing the non-human interests of the ecosystem itself. Generally, the presence of this critical "stakeholder" is represented by the inclusion of technical and professional capacities of NGOs or university scientists, whose role is to provide objective, scientific assessment of environmental values and services being provided. Although this is, no doubt, a useful step, it fails to take account of the (usually greater) knowledge of indigenous people.

In Austria, at the core of mountain policy is a requirement to undertake valuation of non-marketable goods, which must be included in comprehensive assessments aiming at sustainable development. Accordingly,

The emphasis on the potential local and regional amenity character of mountain areas has made it possible to enhance small-scale development initiatives at the local level. [Thus] sustainable resource use in peripheral mountain regions largely depends on the possibilities ... of including amenities as development potentials in regional concepts, of nurturing the endogenous potential of the local population and of inducing appropriate initiatives for a balanced development of cultural landscapes and rural society. (Dax 2002)

These partnerships – whether between upstream and downstream dwellers, governments and private organizations, producers and con-

sumers, or global communities and local institutions – are quite often initiated by the stakeholders themselves. Incentives for individuals to act
collectively rather than independently encourage stakeholders to continue returning to the table to renegotiate fragile and tenuous partnerships and alliances.

Key actions

The foregoing suggests that the specific types of decision-making power
and participation needed depend on whether the challenge is to create
and/or strengthen local sustainability or to promote linked sustainability.

For governments, donors, and NGOs, actions to promote appropriate
and adapted institutions for sustainable management of mountain resources, critical for both upstream and downstream populations, should
include the following:

- Expanding recognition of the importance and value of environmental
  services to individual and community well-being, regardless of whether
  the services can be marketed.
- Involving local people in resource management wherever possible, and
  paying them sufficiently for their services.
- Restructuring property rights to recognize and encourage community-
  property systems, where appropriate.
- Providing institutional mechanisms for structuring and enforcing stewardship agreements and encouraging their enactment.
- Strengthening procedures to ensure that local people receive adequate
  compensation for the exploitation of resources in their areas by others.
- Improving methods for valuing environmental services, so that stewardship and environmental-mitigation agreements for compensation
  can be reached on market-based principles, where possible.
- Establishing appropriate funds to pay for stewardship services, especially for global environment services and cases where the benefits are
  too diffuse for market-based mechanisms.

Finally, it is important that data be spatially disaggregated at a sufficiently fine level of detail and mapped. Only by identifying the distribution of poverty, the value of environmental services and natural resources, and the extent or potential for market linkages, can appropriate,
targeted, and cost-effective interventions be designed and implemented.
In each case, decentralized and democratic institutions are critical to
sustainability in mountains, whether sustainability is narrowly defined
to encompass self-sufficient local communities, or whether it refers to
linked sustainability where both upstream and downstream beneficiaries
are involved.

REFERENCES

Baral, Jagadish Chandra. 2002. "Unintended outcomes of community forestry in Nepal." Mountain Forum: Contribution to BGMS e-conference, March.

Benavides, Hugo Rodriguez, Jaime Villena Soria, Samuel Ordonez Colque, and Marco Fernandez Valdivia. 2002. "The *Waru Waru*: An alternative technology for the sustainable agriculture in Puno Peru." Peru: PIWANDES Institute of Technological Innovation and Promotion of Development. Contribution to BGMS e-conference, March.

Bhandary, Uddhab K. 2002. "Rural poverty analysis and mapping in Nepal." Abstract of paper presented at the International Seminar on Mountains, RONAST, Kathmandu, Nepal, March.

Brown, Aleta. 1998. "When Push Comes to Shove on the Biobío." *World Rivers Review* Vol. 13, No. 4, August.

Byers, Elizabeth, and Meeta Sainju. 1994. "Mountain ecosystems and women: Opportunities for sustainable development and conservation." *Mountain Research and Development* Vol. 14, No. 3, pp. 213–228.

Chomitz, Kenneth M., E. Brenes, and L. Constantino. 1998. "Financing environmental services: The Costa Rica experience and its implications." Paper prepared for Environmentally and Socially Sustainable Development, Latin America and Caribbean Region, World Bank. Washington: World Bank.

Crawford, David. 2002. "The salience of local labor organization in Morocco's High Atlas." Contribution to BGMS e-conference, March.

Dax, Thomas. 2002. "Endogenous development in Austria's mountain regions: From a source of irritation to a mainstream movement." Mountain Forum: Contribution to BGMS e-conference, March. Based on case study published in *Mountain Research and Development* Vol. 21, No. 2, August 2001.

Duzgun, Mevlut. 2002. "Forestry-based contributions to the sustainable livelihoods of forest-dependent mountain communities in Turkey." Ministry of Forestry, Turkey: Contribution to BGMS e-conference, March.

Hernandez, Arrufo Alcantara. 2002. "Tradition and Conflict in the Organization of Rural Space in the Andes." Contribution to BGMS e-conference, March.

Latigue, Laura. 2002. "Protecting Forests through Co-Management." USAID: Contribution to BGMS e-conference, March.

Lopez, Andres Felipe Betancourth. 2002. "The phenomenon of peasant mobility as a search for territories of peace in the Central Mountains of Columbia." Universidad de Caldas, Columbia: Contribution to BGMS e-conference, March.

Mountain Institute. 1998. *Sacred mountains and environmental conservation.* Franklin, West Virginia: The Mountain Institute.

Mphale. 2002. "Energy demand, access, supply and implications for Lesotho's mountain areas." National University of Lesotho: Contribution to BGMS e-conference, March.

Poteete, A. and E. Ostrom. 2002. "An institutional approach to the study of forest resources." In: J. Poulsen (ed.) *Human impacts on tropical forest biodiversity and genetic resources.* New York: CABI.

Pratt, D. Jane, and John D. Shilling. 2003. "High time for mountains: A program for sustaining mountain resources and livelihoods." Background Paper for *World Development Report 2003: Sustainable Development in a Dynamic World*. Washington, D.C.: World Bank.

Preston, L. (ed.). 1997. "Investing in mountains." Franklin, West Virginia: The Mountain Institute.

Reetz, Geene. 2002. "Water quality: A case study in the San Juan Mountains." US EPA: Contribution to BGMS e-conference, March. Based on case study presented at the National Mountain Conference: "Stewardship and human powered recreation for the new century." Golden Colorado, September 2000.

Stevens, Mervin. 2002. Comment on BGMS-C1, Mountain Forum, March.

Timsina, Netra Prasad. 2002. "Empowerment *vis à vis* marginalization: a mixed reponse from the decentralized institutions for sustainable resource management." Mountain Forum: Contribution to BGMS e-conference, March.

Tupayachi, Epifanio Baca. 2002. "Experiences of local management, Ayacucho, Peru." Contribution to BGMS e-conference, March.

Upadhaya, Madhukar. n.d. "Greening hills: Seeing the forest and the trees." In: *Tough terrain: Media reports on mountain issues*. Asia Pacific Mountain Network and Panos Institute South Asia.

Zandstra, Hubert. 2001. "Integrated management of basins." Presentation by International Potato Center at the International Workshop on Mountainous Ecosystems: A future vision. Cusco, Peru, April.

8

Conflict and peace in mountain societies

S. Frederick Starr

Summary

With few exceptions, the most numerous and obdurate conflicts in the world today occur in mountain zones. Although each presents specific features, it is possible to speak of a generalized problem of social and economic breakdown in mountain regions almost everywhere. Caught between isolation and integration, oppressed by indifferent or ineffective governments, yet with enough access to modern communications to know that they are being slighted, mountain people resort to desperate measures. Drug production, a psychology of victimhood, and the lure of radical movements are the outward manifestations of social and economic breakdown.

The only effective way to resolve these conflicts is to address the social and economic issues that provide their seed-beds. This means combining initiatives fostering security and economic development. It means developing the skills that enable people to feed themselves and their families and to create jobs. It means focusing on people rather than things.

To successfully resolve mountain conflicts, the international community must begin systematically to monitor economic and social conditions in the affected regions. Because "we do what we can measure," this is an essential precondition for effective action on this critical issue.

Conflicts in mountain zones: The issue

For most of the past half-millennium, the main source of conflict in
mountain regions has been the effort of emerging states to extend their
power over mountain peoples. To a greater or lesser degree, wars in
Scotland, Switzerland, Peru, the North Caucasus, Afghanistan, and large
parts of Mexico all arose from this process. Once the issue of modern
states' power over upland peoples was settled, however, most mountain
zones faded from world attention.

Recently, a fresh wave of conflict has swept the mountain regions of
several continents. Within the short span of the past decade, places as
diverse and distant from one another as the Peruvian Andes, the Bal-
kans, the Afghan Hindu Kush, the Nepalese Himalaya, Karabakh and
Chechnya in the Caucasus, the Colombian highlands, the Atlas Mountains of Algeria, Rwanda/Burundi and Ethiopia/Eritrea in Africa, and
the Pamirs of Tajikistan have all witnessed bloody fighting. Hundreds of
thousands have died in these struggles.

These conflicts at high altitude undermine the conditions that sustain
human life in these fragile zones. Although fought mainly by men, they
define the conditions under which women and children struggle to survive. Where they might once have been waged in obscurity, now such
wars are reported throughout the world and in such a way as to draw
major powers into their vortex. The European Union, India, Russia, and
the United States have all found themselves engaged in military operations at or above the timberline. Because the precipitating issues are
never solely local in their import, conflicts in remote mountain areas
readily become sources of tension in the larger community of nations. It
is no exaggeration to say that the problem of war and peace in mountain
areas is among the most urgent and intractable issues of international
relations today.

Until recently, few were prepared to acknowledge the existence of a
"mountain problem" as such. Even today it is convenient to treat each
instance of armed combat in mountain areas as unique. Those who take
this course often trace the roots of each instance of conflict to age-old
local ethnic or religious tensions (see Case 1).

Case 1. Age-old conflicts in Rwanda: the ethnic factor

Even age-old ethnic conflicts have causes. The 90-year history of conflict between
the Tutsi and Hutu tribes of Rwanda, for example, traces less to the fact that the
former were pastoralist and the latter agricultural than to the fact that German
and Belgian colonial rule favoured the Tutsis and institutionalized discrimination
against the Hutus. A Hutu revolt in 1959 led to Rwandan independence in 1962

but fuelled rather than mollified the tensions that already existed. Other exacerbating factors include a scarcity of available land, a rapid fall in coffee prices, Rwanda's geographical isolation that hindered economic development and diversification, and an insecure government that proved capable only of responding reactively. Together, these and other elements created an unstable and violent situation that exploded into gruesome genocide when President Juvenal Habyarimana's plane crashed near the capital of Kigali in April 1994. Pure ethnic tension certainly existed, but this had been heightened and intensified by a host of quite specific developments over nearly a century. Sources: Byers (2002); Percival and Homer-Dixon (1995).

It is true that such factors frequently play a role, and it cannot be doubted that isolated mountain societies easily develop a spirit of lawlessness, if not a sense of persecution. Yet in almost every case where these factors are invoked, the same warring parties and groups managed to coexist with one another for decades, or even centuries, prior to the recent outbreaks. And why, one might ask, should the ethnic or religious identity of a Peruvian Indian and an Afghan Pashtun suddenly become an issue at nearly the same time, and why should armed Nepalese peasants manifest the same forms of resistance when those against whom they fight are of the same religion and ethnicity as themselves?

What do we know about these conflicts? Some common features of mountain-based conflicts

Notwithstanding these unacknowledged and unresolved problems of analysis, recent scholarship has produced an impressive body of research relevant to problems of war and peace in mountain settings. Anthropologists, for example, have studied both the origins of specific mountain conflicts and the traditional means of conflict resolution through which the participants seek to settle them. Sociologists have traced the weakening of communal bonds under the impact of external forces, and the manner in which armed struggle might either foster greater cohesion or lead to social collapse. Economists have examined the decay of mountain-village economies as they are sucked into the whirlwind generated by emerging urban-centred systems of resource use, production, and distribution. Political scientists and historians have followed the way in which armed strife in the most distant mountain settings evolve into national and then international crises of the first order.

In light of this, is it appropriate to speak of a generalized problem of war and peace in mountain settings today? Or must we, instead, continue to treat each instance of mountain-based conflict as unique, and the fact of the simultaneous appearance of more than a dozen of them on three continents as merely coincidental? Let us first acknowledge that the conditions that precipitate conflict and mobilize support for armed action among mountain peoples, and between them and outsiders, are always highly specific. The concerns of native peoples in the hills of Chiappas, the goals they seek, and the way they choose to pursue them all differ sharply from the corresponding issues in Burundi, Bosnia, or Tajikistan. Yet, this said, it is possible to draw from the many mountain conflicts of the past decade a number of elements that are common to most of them, if not universal.

On the dangerous border between isolation and integration

An important commonality among conflict-prone mountain regions is that they are neither totally isolated from the modern world economy nor fully integrated into it. Their communications and transportation infrastructures are sufficient to enable national businesses or markets to exploit mountain resources, whether gold in Kyrgyzstan or amber in Chiappas. Both managerial and financial control of this process lies elsewhere, however. As a result, mountain peoples reap little benefit from their involuntary participation in world markets.

To take but one example, profits from the sale of hydroelectric power from mountain areas invariably flow into metropolitan coffers. In addition, whereas oil- and gas-rich desert countries can demand payment for their God-given resources, mountain people are condemned to sending out their precious water without charge. Is it any wonder that among the most conflict-prone communities are those which export natural resources to the metropolis but are unable to purchase essential goods in return?

A distinctively modern type of poverty

Poverty has long been a feature of life in many high-altitude communities. However, the poverty that prevails in many mountain areas today is of a peculiarly modern sort, in that it arises from a growing dependence on lowland metropolitan centres rather than from age-old self-sufficiency in a harsh environment.

The ineffectiveness of governments

Governments are not blind to the existence of such poverty, but its alleviation is rarely a high priority. Mountains are often distant from the capital and from main centres of population. Many mountain ranges mark national borders and are therefore treated as security zones. Remote, and subject to "subversive" influences from abroad, they are easily ignored. This is all the easier because the voices of mountain people are rarely audible in parliaments or governmental agencies. As a result, central governments pay little or no political price for subordinating their needs to the more urgent demands of the heavily urbanized regions.

Strategies of development pursued by both socialist and capitalist states during the last century provided theoretical support for policies that ignored mountain peoples. Extensive modes of development placed great emphasis on economies of scale of a sort that could be achieved only through industrialized agriculture or in major urban factories. As long as these strategies held sway, governments viewed mountain settlements as little more than sources of inexpensive labour for large enterprises elsewhere. Officials responded to mountain poverty by suggesting that its victims migrate to the lowlands; in the Soviet Union they went further, forcefully resettling whole mountain nations in large lowland collectives.

In those rare cases where central governments acknowledge the problem of mountain poverty and seek to do something about it, they generally have no idea how to proceed and lack the necessary financial resources for doing so. Nevertheless, the mere fact that national governments claim to rule in the name of their entire population politicizes the problem of mountain poverty. Mountain residents see the unwillingness or inability of central governments to alleviate their poverty as proof that the state has abandoned them in their hour of need.

Communications create self-awareness among the mountain poor

The steady advance of modern communications technologies into even the most remote areas deepens the alienation of mountain peoples from the national polity. Radio and television enable even illiterate subsistence farmers and miners at remote facilities in the mountains to form some conception of life in the burgeoning lowland cities and in their national capital. They become acutely conscious of their relative backwardness, even if their situation is improving in absolute terms. They send their brightest sons to seek jobs in the cities in the hope that they will remit money to their parents and those left behind. With many of their best and brightest members departing for jobs in the lowlands, mountain societies slip ever further behind, and are daily reminded of this fact by reports and images of urban life beamed to them over the electronic media.

Desperate remedies: Drugs as "globalization for the poor"

As despair deepens, mountain people are prepared to abandon their traditional occupations in favour of whatever activities will enable them to feed themselves and their families. In many cases this means the cultivation of opium poppies or coca plants. This puts them into contact with criminal elements from their own communities, the capital cities, and foreign lands. Gradually, they are drawn into the lowest ranks of this most dangerous form of economic globalization. For mountain people, the narco-industry is a kind of "globalization for the poor" (see Case 2).

Case 2. The values of coca

Involvement with drug production often combines tradition and modernity in complex ways that those seeking its eradication are disinclined to acknowledge. The use of coca leaves has been part of the lives of Andean peoples for centuries, strengthening community practices among Quechua, Aymara, and Amazonian groups in Brazil, Colombia, Ecuador, and elsewhere. Valued for its nutritional qualities, the coca leaf also plays a role in the larger world view of many communities. Meanwhile, in both the cities of the Altiplano and the valleys, the use of coca leaf has spread to the middle and upper classes, notably in Bolivia. This provides a link to the larger world. When structural adjustments are introduced in the national agrarian economy, traditional sectors of production decline, displacing large numbers of people, many of whom turn to drug production in order to survive. Efforts at total eradication are built on a "dialogue of the deaf," in which the relationship of local people to both tradition and necessity is largely ignored. Source: Delgado (2002).

A psychology of victimhood

The deepening poverty of mountain peoples, whether absolute or relative, combined with their growing consciousness of their fate, gives rise to a highly volatile psychology of victimization. Mountain folk who have lived at peace for generations suddenly start to lash out at their supposed oppressors. Most of their targets are local – more-prosperous residents of a neighbouring valley, or a nearby ethnic group whom they suspect of conspiring against them with merchants or officials. As the struggle over scarce resources deepens, conflicts can arise even among rival family groups or clans within a single community. In Dagestan, for example, where unemployment reached nearly 80 per cent in the rural mountainous areas, leaders of several ethnic groups formed political movements to support their claims to status and resource sharing.

Radical non-governmental initiatives

The failure of governments to address these problems leaves the field open to non-governmental forces. The breakdown of legitimate authority in mountain zones gives rise to local warlords, criminal groupings who extract money from the population in exchange for minimal security (see Case 3).

Case 3. Sources of conflict in Colombia

Conflict gives rise to conflict, as can be seen in the central mountains of Colombia. There, some 1.5 million people have been displaced by conflict over the past decade. In their search for "territories of peace" these migrants, nearly half of whom are minors, transmit instability to new areas and provide a steady flow of recruits, both voluntary and involuntary, to armed groups. Indeed, pressure caused by recruitment is one of the causes of displacement. At the same time, the depopulation of areas from which migrants fled creates a kind of vacuum that generates new sources of conflict. The Colombian Constitutional Court issued a 1997 law intended to protect displaced populations, but this has proved to be a dead letter in practice. Instead, communities defined in terms of "adherence to a place" grow ever weaker and a range of sociological and psychological pathologies set in. These are promptly exported to urban areas, which are then swept up in the same conflicts from which the migrants fled. Source: Lopez (2002).

Warlords, in fact, provide what might be called surrogate governmental services in the absence of the real thing. While some of these home-grown warlords are coldly exploitative (e.g. Bosnia's Nasir Oric), others assume the role of Robin Hood, championing the poor against officialdom or other presumed oppressors. All too easily, though, such bands forge links with narcotic traffickers, thus further criminalizing mountain society: one such example is Afghanistan's Rashid Dostum.

International support and funding

These conditions provide fertile soil for the spread of extremist ideologies and movements arising from distant urban areas or foreign lands. Whether secular or religious, such movements invariably seek to mobilize the oppressed in the name of a radical transformation of all society. As the result of money provided by sympathizers abroad or deriving from illegal activities, including narco-business, these movements can offer generous financial support for their recruits in the form of free training, aid for families, and employment.

Government responses that make matters worse

Finally, even the most distant and aloof central governments are forced to react to the existence of armed conflicts in their countries' mountain regions. However, their first response invariably comes late in the day, long after the normal functions of civil society have broken down. Hence, central governments treat the issue primarily as a security problem. The army is sent in to quell the unrest, but it immediately becomes part of the problem it was sent to resolve. The national army introduces ever more potent armaments to the battlefield. As these fall into the hands of the other combatants, the entire mountain zone is progressively militarized and the army becomes just one fighting force among many. In the end, social and economic breakdown is complete and becomes all but irreversible.

These, then, are some of the common features of conflicts arising in the world's mountain regions in the early years of the new millennium. This social morphology is subject to many local variations, but its overall pattern and direction is everywhere the same – from the Andes to the Himalayas, from the Atlas Mountains to the Caucasus. Not one of these diagnostic traits is absolutely new in our world today: examples of most of them can be found in earlier conflicts in the nineteenth and twentieth centuries. What is distinctive to our era is that all of these features now regularly occur together, and in a compressed time period, which renders them all the more potent.

Possible remedies

What, if anything, can be done to break this cycle of despair, civic collapse, and conflict? Are there "best practices" that can head off this vicious cycle, or break it once it has begun? The record of recent years is not encouraging; none the less, a few obvious prescriptions can be cited.

Mountain conflicts as an international issue

Mountain-based conflicts must no longer be considered as purely domestic affairs. Few mountain conflict zones are confined neatly within the borders of one country. Moreover, the frequent importation of foreign arms, the participation of foreign fighters, the involvement of international drug cartels, and the exploitation of mountain conflicts by neighbouring states all underscore the extent to which they must be treated as international issues. As such, these conflicts and the conditions leading to them must become the subject of international, as well as national, consultation and action.

Monitor social and economic conditions in mountain regions

For such initiatives to be effective at either level, they must be based on reliable data on the actual conditions in mountain regions. These data do not currently exist. The United Nations' Human Development reports and other such statistical overviews should henceforth include a category dealing with each nation's mountain territories and peoples. The economy, ecological conditions, state of human welfare, and public health in mountain regions should all be disaggregated from the national whole for analytic purposes. As the proverb says, "We do what we measure." At present, the living conditions of people in the world's mountain zones are not being adequately monitored.

Focus energy on specific cases to prove that success is possible

International institutions dealing with peacemaking and peacekeeping must acknowledge the unprecedented global proliferation of conflicts in high-altitude zones and focus new energy on reversing this trend. In the absence of clear successes, policy makers will persist in the common view that mountain conflicts are, *ipso facto*, intractable and can therefore only be fenced off rather than resolved.

Identify "best practices" that have produced success

Any long-term successes in averting or resolving such conflicts must be identified and their basic elements set forth, in order to foster a "best practices" approach to mountain societies and their problems (see Case 4).

Case 4. Learning from tradition

"Best practices" may derive from the traditional means employed by mountain people themselves to resolve conflict. A vivid case in point is the institution of *Chhinga* employed by the Jaunsari tribal community of Uttaranchal State in the Himalayan mountains of India. When irresoluble disagreements or conflicts arise between two families, rather than fight they engage in *chhinga* or *varjan*, under which the contending families will not enter each other's houses, eat together, or share the water pipe (hookah). This condition continues until one of the parties is prepared to accept that it is at fault, which generally comes when it feels the wrath of the supernatural power that was appointed as the keeper of the *chhinga*. This may take months, years, or generations. During this time each party continues its full role in community life and even collaborates with the other to the extent local custom requires, while maintaining social distance with the opposing side. Source: Joshi (2002).

Hitherto, the prevailing approach everywhere has been improvisational rather than systematic. Only through the development of a comparative understanding of the problems and pathologies of mountain communities can more general remedies be devised, applied, and improved on the basis of experience.

Address security issues and social/economic problems together, not separately

The tendency everywhere today is to address the political and military issues first, and only then to turn attention to the social and economic problems that gave rise to conflict in the first place. Henceforth, both must be addressed simultaneously and from the very outset. Those who object that economic and social development can take place only after peace is established should study the several instances where quiet economic development at the village level has led to the withdrawal of whole families, groups, and communities from the prevailing conflict.

Focus remedial measures on people, not things

In both resolving existing conflicts and averting future ones, the emphasis should be placed on policies that deal with people rather than things. Large-scale infrastructure projects should not be excluded, but the benefits they bring are rarely commensurate with their cost. By contrast, initiatives of any scale that are framed in terms of the actual needs of mountain communities are most likely to be cost-effective. The experience of several of the leading NGOs, from Pakistan to Peru, suggests that the most productive of these initiatives are often quite modest in scale and cost, focusing on village-level agriculture (see the mandate of the International Centre for Integrated Mountain Development (ICIMOD) [http://www.icimod.org/general/brochure1.htm]).

Actively engage mountain people themselves

Progress towards stabilizing mountain societies and averting future conflicts will require that mountain people be actively engaged as participants in, rather than treated as passive recipients of, international assistance. Stated differently, national and international agencies must engage local people in such a way that they willingly accept their share of responsibility for the success of any given programme. Although democratic participation will not itself transform life in mountain areas, it is surely better to ask mountain dwellers about their needs and interests than simply to tell them (see Case 5).

Case 5. Engaging local people in alleviating conflict

The case of Nepal vividly demonstrates that there is really no workable alternative to engaging the local population in the alleviation of conflict. There, numerous "projects" by well-intentioned national and international groups have led to unanticipated results. Misappropriation, rent-seeking extortion, other forms of outright corruption, and the confrontation of reformist legal norms with local customary laws and practices undercut effectiveness at every level. Worse, hostility engendered by what indigenous people see as the indifference, arrogance, and even corruption of project personnel discredits the efforts of honest and well-intentioned staff members and undermines whatever trust the few honest and scrupulous bureaucrats have managed to foster. Under such circumstances, engendering trust (which can be achieved only through the participation of local people as equal partners) becomes the *sine qua non* for all effective action. Source: Upreti (2002).

Embrace the private sector and market mechanisms

Initiatives directed towards long-term economic and social development in mountain communities must fully embrace the private sector, fostering initiative and entrepreneurship under conditions of a market economy. This means recognizing that most mountain peoples aspire to participate fully in modern life, even while maintaining their connection with their traditional homes and cultures. Failure to apply this simple rule will further confine mountain communities to ghettos within their national and regional economies.

Nor should development deny to mountain people the access to regional transport and trade that other citizens enjoy. The urge of many well-intentioned experts to seek to "protect" mountain communities from the worst effects of modernity all too easily becomes a rationalization for non-development, especially in the transportation sector. Mountain people should have the same right to choice in the area of transport and trade as their fellow-citizens in the lowlands.

Help mountain people become protectors of their environment and resources

Such initiatives must be designed and carried out in such a way that they help local residents to acknowledge the special fragility of fauna and flora at high altitudes, as well as the finite character of non-renewable resources. Failure to do so will destroy the resources that will make progress sustainable.

Provide education and training that build skills for economic and social development

Because the only people who, in the long run, can prevent or resolve conflicts in mountain regions are the local residents themselves, they must have access to all the various forms of education and training

that will increase their effectiveness at these tasks. With only a few notable exceptions, such programmes for building modern capacities among
mountain peoples do not exist today; however, through the development of new types of institutions calling on the latest technologies, such
education and training could be delivered to even the most remote and
conflict-bound regions worldwide. Compared with the human and economic toll of conflict, the cost of such innovative initiatives in education
and training is modest.

REFERENCES

Byers, A.C. 2002. "Ethnic conflict in Rwanda." Case Study. Mountain Forum e-
consultation for the Bishkek Global Mountain Summit, UNEP.
Delgado, R. 2002. "The cultivation of coca in Bolivia: Symbol of life and death."
Case Study. Mountain Forum e-consultation for the Bishkek Global Mountain
Summit, UNEP.
Homer-Dixon, T., and V. Percival. 1995. "Environmental scarcity and violent
conflict: The case of Rwanda." Occasional Paper for the Project on Environ-
ment, Population, and Security. Washington: American Association for the
Advancement of Science and University of Toronto.
Joshi, P.C. 2002. "Chhinga as an institution of suspension of conflict." Case
Study. Mountain Forum e-consultation for the Bishkek Global Mountain Sum-
mit, UNEP.
Lopez, A.F. Betancourth. 2002. Grupo ASPA, Universidad de Caldas Instituto de
Educacion Superior Colegio Integrado Nacional Oriente de Caldas. "The phe-
nomenon of peasant mobility as a search for territories of peace in the Central
Mountains of Colombia." Case Study. Mountain Forum e-consultation for the
Bishkek Global Mountain Summit, UNEP.
Upreti, B.R. 2002. *Management of social and natural resource conflict in Nepal:
Reality and alternatives*. New Delhi: Adroit Publishers.

9

National policies and institutions for sustainable mountain development

Annie Villeneuve, Thomas Hofer, Douglas McGuire,
Maho Sato, and Ali Mekouar

Summary

To implement sustainable mountain development, the creation of national policies and institutions that address the specific needs of mountain regions is very important. Facing harsh socio-economic conditions (including poverty, isolation, and lack of infrastructures), and being characterized by special environmental features (such as climate, hazards, and erosion), these regions require a differentiated approach to sustainable development. This chapter provides an overview of the different policy approaches and institutional options that currently exist in a number of countries to further sustainable development of their mountain regions.

For both policies and institutions, two distinct categories can be identified: (1) countries where mountain-specific policies/institutions have been created; (2) countries where mountain issues have been taken into account through existing policies/institutions.

First, this chapter addresses policy approaches. Integrated and participatory approaches to sustainable mountain development have been adopted in some countries. Austria, France, Georgia, Poland, and Cuba have adopted specific mountain policies that respect the particular needs of mountain areas and peoples. Morocco, recognizing the difficult socio-economic conditions prevailing in its mountain regions, is developing a mountain-specific policy. Where national mountain-specific policy does not exist, measures concerning mountains are added to existing sectoral policies to provide certain orientations for mountain development. Nepal and Japan, both well known for their mountainous regions, use sectoral policies to achieve sustainable mountain development, as does Italy; Bulgaria and Colombia, on the other hand, focus on mountain protection through policies regarding protected areas.

The importance of adequate institutional mechanisms for effective policy implementation is highlighted. In some countries, institutions that specifically deal with mountains have been created to carry out mountain policies. Sometimes, governmental institutions are established to act as the sole or principal structure in charge of mountains: France and Viet Nam have successfully created such mountain institutions. NGOs and other associations are increasingly involved in the implementation of national policies in mountain regions: Georgia, Tanzania, and Switzerland, for example, have NGOs active in sustainable mountain development. The role of the IYM national committees is also briefly described. Most countries, however, have not created specific institutions to deal with mountains and prefer to rely on existing institutions to perform mountain-related functions. Croatia and Mexico provide examples of a good institutional coordination between relevant ministries, and the Philippines has established a policy-making and coordination body for a mountain forest reserve that deals specifically with mountain issues.

Recommendations are made to enhance sustainable mountain development through appropriate policies and institutional frameworks. A few illustrations of the kind of activities required in this regard are presented.

Introduction

One objective of the International Year of Mountains (IYM) was to "guarantee the present and future of mountain communities by promoting the conservation and sustainable development of mountain regions" (FAO 2000). The goal of sustainable development is to create and maintain a permanent balance between human and natural environments by dealing with economic, social, and environmental issues on an equal basis. The creation of policies and institutions at the national level that meet the specific needs of the mountain regions contributes in a decisive manner to the implementation of this goal (see below).

Sustainable-development challenges that are linked to national policies and institutions

Mountain regions have special characteristics (geographical isolation, political marginality, difficult climatic and environmental conditions, fragile ecosystems) that militate in favour of a differentiated approach to sustainable development.

Mountain regions, however, over and above their general characteristics, differ considerably from one another. A precise definition of sustainable mountain development is not conceivable (Price and Kim 1999), inasmuch as a regional approach must prevail over a global approach in this field. The principal challenges of this development remain identical, nevertheless, in their dual dimension – socio-economic and environmental (Lynch and Maggio 2000).

Socio-economic dimension

The characteristic hemming-in of mountain regions results in isolation, poverty, and (frequently) precarious living conditions for their inhabitants. Inadequate, or often non-existent, infrastructures accentuate the gulf that separates them from the more advantageous conditions of the open-plain regions. These handicaps must be recognized. The effective participation of mountain communities in making decisions and implementing mountain-development actions is indispensable, owing to the invaluable knowledge represented in their traditional practices and indigenous experience. Although they often have traditional and indigenous tenure systems, mountain communities are frequently not granted formal property rights. Developing the potential that mountain regions contain (in agriculture, forestry resources, tourism, and local products) is a paramount objective of sustainable mountain development.

Environmental dimension

Mountains represent vital ecosystems that contain the world's principal water reserves as well as rich plant and animal diversity. However, unfavourable climatic conditions, hazards (such as avalanches and landslides), erosion, and soil infertility are alarming factors that urgently call for increasing public awareness in this regard. Measures for safeguarding the mountains must, therefore, be made an intrinsic part of mountain-area development programmes, in order to guarantee the proper management of natural resources. These challenges vary from one region to another, particularly as a result of environmental characteristics (e.g. climatic conditions) and human activities (e.g. deforestation). It is, therefore, the role of central governments and decentralized authorities to define the best practices for managing fragile resources and sustainable mountain development (FAO 2000).

This chapter provides insight into the alternative political approaches and institutional options that have been created in a number of countries in order to further sustainable mountain-region development. Given its brevity, it is far from exhaustive, since only a limited (although representative) number of national experiences in this field are dealt with, and the policies and institutions touched upon as examples are not examined in full detail. A longer report (Villeneuve, Castelein, and Mekouar 2002) covers many of these issues in greater detail.

The policies and institutions that are briefly described here can be grouped together in two clearly separate categories:

- those that have been specially created in order to meet the particular needs of mountain regions;
- those that relate to one or another economic development and/or environmental-protection sector (such as agriculture, forestry, tourism, biodiversity, and territorial development) involving the mountains and that serve as a more or less suitable political and institutional framework in this regard.

Policy approaches

The policies that apply to areas with strong physical constraints such as the mountains (e.g. climate, relief, altitude) must be considered in light of the sectoral policies dealing with such factors as agriculture, forests, urbanism, transport, and education, which affect entire national territories. The appropriate path to follow involves an integrated and participative approach to sustainable mountain development that would make it possible to harmonize sectoral policies and direct them towards the same objective – the protection and development of mountain regions (FAO and Italy Cooperative Programme 2000). Before adopting a new set of policies, however, a detailed study must be made of the costs and advantages that would result from implementing them. This decisive step will dictate the type of action to be taken – that is, whether (or not) to create an ad hoc policy for the mountains.

In a number of countries, however, the question has not been put in these exact terms – or, when it was, the answer was negative. In the absence of mountain-specific policies, the development and conservation of mountain areas are subjected to the orientations of the relevant sectoral policies, including those regarding such factors as forestry, agriculture, tourism, territorial development, and rural development. Specific and related measures have occasionally been implemented in order to adapt these policies to the characteristic traits of mountain regions. An approach of this nature can prove to be advantageous in practice, as it would be based upon policies that have been well tested and benefit from substantial financial support. Although this policy option has been subjected to repeated criticism (FAO 2001), it would not automatically lead to failure merely because it is not explicitly confined to mountain regions. By remaining in touch with the national context, it might possibly constitute an appropriate platform for sustainable mountain development.

Specific policies for sustainable mountain development

The following examples of national policies that are specific to mountain areas indicate the different forms that such policies can assume.

Austria: A multi-sectoral mountain policy

As mountain regions cover more than half of Austria, regional development, agriculture, and the protection of mountain forests have, for a number of years, been among the principal concerns of the country's economic and territorial development. Beginning in the early 1970s, a Special Programme for Mountain Farmers emphasized the different functions carried out by agricultural activities that affected mountain people as well as territorial conservation (Dax 2002). The government declared in 1996 that the preservation of mountain agriculture and forests was one of the principal national priorities (Lynch and Maggio 2000). The objective of Austria's mountain policy is to create better living conditions for the country's population, to preserve the mountains' environmental wealth, and to increase its resources by promoting regional development. The underlying principle guiding Austria's national mountain policy is the diversification of the mountains' assets – forest, agricultural, tourist, and environmental.

France: An evolving policy regarding mountain ranges

French policy makers have been involved with mountain issues since the 1960s, when the government promoted agricultural and territorial-development policies that were based upon the interests of rural populations. Their mountain-regions policy therefore originated as a cross between agricultural policy and regional-development policy. Mountains were officially recognized in the national and regional development directive regarding the protection and development of mountain regions (Decree No. 77-1281 of 22 November 1977) as separate entities that combined ecological, agricultural, forestry, and tourism functions. France's national mountain policy, which was institutionalized in 1985 with the implementation of the law relating to the mountain regions, aims to "allow local populations and their elected officials to acquire the means and control of their development with a view to creating equality in terms of income and living conditions between the mountain and other regions in full respect of the mountain peoples' cultural identity" (Article 1). This policy, which receives financial support from both the French Government and the European Union, was first created in order to deal with the handicaps that characterize mountain regions, including natural constraints (high altitudes and temperature changes) as well as land ownership and logistical constraints (Villaret 1996). Current implementation of the government's policy is directed at the development of the assets and potential of each mountain range (massif) in accordance with its special characteristics in terms of goods and wealth (such as culture, nature, landscape quality and products, leisure activities, and sports). This policy with regard to the country's ranges was confirmed in 2000 as France's integrated interregional and interministerial policy, with the objective of strengthening the legal authority of prefects in their role of coordinators for specific ranges (Ministère de l'aménagement du territoire et de l'environnement [Ministry for National and Regional Development and the Environment] 2001).

Georgia: A mountain policy to be implemented

Georgia's mountain regions cover over two-thirds of the country and are vital to the national economy, in particular because of their significant contribution to agricultural production, the beauty of their landscapes, and the wealth of their recreational and tourist potential (European Mountain Conference 1998). Given Georgia's particular condition as a country in transition (including the after-effects of the civil war, natural catastrophes, and the economic crisis), the socio-economic conditions of the mountain regions are more precarious because the natural environment has also been seriously affected (by such factors as soil erosion, deforestation, the degradation of cultivable lands and pastures, and the loss of biodiversity). Faced with this alarming situation, the Georgian Constitution provided for the possibility of using the legislative process to create special privileges "in order to guarantee the economic and social progress of the high mountain regions." This unique constitutional measure is evidence of the central role played by the mountain regions in Georgia's national landscape, confirmed in June 1999 by the adoption of the law regarding the socio-economic and cultural development of the mountain regions. This law reflected the country's desire to enhance the importance of its mountain regions and provided for the implementation of a "policy for the socio-economic development of mountain regions." The law stipulates measures for encouraging the development of mountain regions and envisages concrete means of achieving this objective. Unfortunately, the implementation of this encouragement policy is adversely affected by the country's current economic difficulties. The government is hoping to obtain financial backing from abroad in order to be able to deal, at least in part, with these obstacles.

Poland: A policy based upon an extensive mountain approach

Poland has created a national policy for the protection of mountain regions in line with the orientations of the country's national environment policy. The priorities that have been set by this mountain policy include, in particular, the protection of nature, landscapes, and waters; the sustainable management of forestry resources; and the development of the recreational potential of the mountain regions. It is interesting to note that the mountain regions are regarded, in an extensive manner, as areas with an altitude of more than 350 metres – an altitude that is far lower than the average set by most other European countries. Using this criterion, approximately 8 per cent of Polish territory is mountainous, resulting in a relatively enlarged area for the application of the country's mountain policy. To support this policy, an Act concerning the economic development of mountainous areas was developed in 2001 but is yet to be formally endorsed. The Act defines criteria for mountain delimitation and would provide assistance measures to farmers and farming communities in mountainous areas. Before this recent legal development, the mountain policy was based on Government Resolution No. 4 of 1985 (Fatyga 2002).

Cuba: An active 15-year mountain policy

Mountains cover nearly 18 per cent of Cuba's land area and host 700,000 inhabitants – 6 per cent of the national population. They represent fragile as well as strategic ecosystems that are essential in terms of both biological diversity and national defence and therefore require a special programme for their conservation and sustainable development. Over the last 40 years, steady efforts have been made to improve the living conditions of mountain communities, mainly through free access to education and health services. In 1987, a Mountain Development Programme, called "Plan Turquino" (after the name of the country's highest peak), was put in place, and its implementation was made a national priority. The plan was designed as an integrated programme for the mountains and could benefit from various government efforts aimed at promoting economic, political, and social changes. Under the programme (among other achievements), numerous education centres at primary, secondary, and superior level were created, and health services were established in most of the country's mountain villages. To carry out the programme, a governmental commission was established by decree, with decentralized branches at the provincial and municipal levels, which allowed the involvement of local communities in the plan's implementation (Perera 2002).

Morocco: A mountain policy that is being developed

From April 1999 to October 2000, the Moroccan Forestry Minister, in collaboration with the government departments concerned, held a series of discussions in order to define a "specific policy for the protection and the development of mountain regions" in Morocco. Three principal reasons favouring the implementation of this policy were thus identified: (1) the isolation and marginalization of mountain regions, which risk the acceleration of rural flight and constitute a source of social instability; (2) soil degradation, as well as that of wooded areas and other mountain resources; and (3) the high development potential of mountain regions and its impact upon the creation of jobs (in agriculture, tourism, handicrafts, and commerce). It also became clear that a specific policy favouring mountain regions and their communities required adapting different sectoral policies (in forestry, agriculture, land ownership, etc.) to the particular needs of mountain areas and populations, based upon an integrated-development approach, as well as the convergence of these policies towards an overall development policy. It was also agreed that the participation of mountain populations should be encouraged through different mountain-development projects. Finally, it was agreed that a mountain policy must provide not only the financial support needed but also the administrative (competent institutions), technical (capacity training), and scientific (education and training) mechanisms that are indispensable for implementing the policy (Meknassi 2002; Ministère chargé des eaux et forêts 2000). In April 2002, the Interministerial Council on Rural Development considered a strategy for the protection and the development of mountain regions. Based on a range approach, the proposed strategy aims at coordinating sectoral policies and programmes, ensuring public participation as well as national solidarity for the benefit of mountain areas. The preparation of a mountain law is under way under the guidance of the above-mentioned Council.

Sectoral policies that affect mountain development

In countries without a national mountain-specific policy (as frequently occurs in practice), existing sectoral policies can provide policy orientations that are more or less explicit – depending upon the country in question – with regard to development choices for the mountain regions concerned. The following are four brief examples of this approach, as employed in Nepal, Japan, Bulgaria, and Colombia.

Nepal: Mountain-oriented policies

Mountains cover more than three-quarters of Nepal and thus constitute a reality that cannot be ignored in all of the country's sectoral policies. Nepal's Community Forestry Programme provides an excellent example of the extent to which its mountain regions are systematically taken into account by the country's political decision makers. This programme recognizes the importance of local knowledge and attempts to use this knowledge through the sustainable development of the forests, one of the country's principal natural resources. The participation of local communities in the management and control of forestry resources is guaranteed by the Forest User Groups that were created for this purpose. These groups benefit from academic and technical training programmes dealing with forestry management and development. Studies have indicated that women actively participate in these programmes and that they are also effective forestry managers (Joshi 1997). This policy of decentralizing forestry responsibilities has had a positive rebound effect on the conservation and development of mountain regions.

Agriculture being one of the country's priority areas, an Agricultural Perspective Plan (APP) was formulated in 1995 with a view "to launching the agricultural sector into a sustainable growth path." The plan points out the difficult farming conditions and envisages different means for agriculture development, including the improvement of agricultural infrastructure and the participation of local people. Through this policy, mountain areas are looked at specifically, as regions with particular features (such as weak structure, altitude, steep slope, or excessive grazing) needing specific measures. Recognizing the differences that exist among mountain regions in terms of agricultural conditions, a recent study called for specific interventions depending on the agricultural conditions that prevail in each mountain district, in order to ensure a well-adapted agricultural development (Pradhan 2002).

Japan: Taking mountains into account in national plans and programmes

Japan does not have a comprehensive policy for its mountain regions. However, since these regions cover much of the country, they are directly considered in different national plans and programmes, particularly those relating to forestry resources, national and regional development, protected areas, and natural habitats. Japan's forests are mostly in its mountain regions and are, therefore, subject to the special measures relating to the mountains. The nation's Forestry Agency supervises the conservation of its mountain forestry resources in order to guarantee comfortable living conditions for the local population. The Agency carries out measures for the prevention of natural catastrophes, as well as ecosystem conservation and forestry development. It is also charged with the implementation of the law regarding the development of mountain villages (the Mountainous Villages Development Act). In 1994, Japan created a national plan for the environment (Basic Environment Plan), with a chapter dealing specifically with the mountains. The plan implicitly recognizes the environmental characteristics that are specific to mountain regions and the effects that they have on the living conditions of local populations. It also provides for the creation of protected areas that are intended to guarantee the conservation of the mountain ecosystems. Finally, Japan is in the process of creating a national ecotourism plan that is intended to favour both the protection of mountain ecosystems and the economic development of regional communities.[1]

Italy: Relevance of land use and forest-related policies

Italy is one of the few countries acknowledging the peculiarity of mountain conditions and needs in its Constitution (1947), on the basis of which, national legislation dealing specifically with mountain regions was subsequently developed. Recently, work was started toward a complete reform of that legislation, with a view to furthering the development and conservation of mountain regions, in the light of lessons learned in regional (European Union) and global policy frameworks (FAO and other UN agencies). Within the Italian Parliament, a group known as the "Friends of the Mountain" (*Amici della Montagna*) has recently put forward five priorities for the design of mountain policies: (1) financial support to reduce economic disadvantages in mountain areas; (2) legal, social, and economic acknowledgement of the value of the activities taking place in mountain areas; (3) improvement of the living conditions in mountain areas through adequate legislation; (4) enhancement of mountain cultural and environmental assets; and (5) devolution to local authorities of the benefits arising from the use of mountain infrastructure.[2] Hitherto, the protection and the development of mountain areas have been taken into account in various sectoral policies. As forest resources are mainly located in mountain areas (95 per cent), public forest policies are highly relevant to these areas. A recent study pointed out that current forest-management and related policies are based on a multifunctional approach, aimed at promoting the protective role of mountain forests (soil protection and water conservation) as well as their productive role (wood production). Likewise, land-use planning tools usually address the needs and constraints of farming development, livestock management, and wood harvesting in mountain areas (FAO 2002).

Bulgaria and Colombia: Impacts on mountains of policies regarding protected areas

A number of countries have implemented policies involving the creation of protected areas (such as national parks and nature reserves) in order to preserve areas of exceptional value from an environmental point of view. All or part of these areas are generally located in mountain regions. In Bulgaria, for example, measures have been taken to protect the habitat as well as the species and characteristics of certain mountain regions by creating national parks and nature reserves in these areas. As a large majority (85 per cent) of these are high-altitude areas, the mountain ecosystems indirectly benefit from these protection measures (Guirova 1995). On the other hand, a Colombian case illustrates some negative social impacts of policies regarding protected areas. In 1997, the government bought an Andean forest reserve ("Selva de Florencia") and placed it under a protection regime for the preservation of that ecosystem and the conservation of its biodiversity: Thus expropriated, native communities have been forced out of the protected area. A survey undertaken four years later concluded that, although there has been an improvement in natural resources, the initiative had generated significant impoverishment of the communities, as many family members had been forced to migrate because of the lack of job opportunities. The survey also highlighted the risk for the protected area if the communities decided to return to the land that they used to occupy (Rivera 2002).

Institutional options

The effective application of sustainable mountain-development policies is dependent upon the existence of suitable institutions that can implement them. The institutional mechanisms required for this vary according to the diversity of the country concerned – whether in terms of administrative systems, economic conditions, social structures, cultural conditions, or financial means, among other distinctive elements. A number of different institutional options are then possible. An appropriate infrastructure for the policies adopted must be created in order to make certain that these are effectively carried out. Two institutional approaches are normally followed, as is the case with questions of policy: either ad hoc structures can be created that meet the particular needs of mountain regions, or optimal use can be made of the structures that already exist (perhaps adjusting them if necessary), in order to rationalize the development of the mountain areas. In both of the above cases, the public institutions and the NGOs often operate in a concrete manner, at the central or command level as well as at the local level.

Institutions that particularly deal with mountains

Governmental institutions

A country may choose to create a new institution as its sole or principal governmental structure charged with sustainable mountain development. A priori, this solution might appear to be ideal. It might, however, lead to conflicts between the newly created specialized structure and the institutions that already exist and have jurisdiction with regard to the mountain regions, of which they would be deprived. Experience in other areas has shown, in effect, that existing administrative bodies are frequently very jealous with regard to their prerogatives and normally resist the intrusion of new institutions rather than offering to collaborate with them. Similar reactions have also been noted with regard to the institutional frameworks created for the integrated management of coastal areas (FAO 1998).

There are, nevertheless, instances of institutional reforms that have succeeded. In France, one of the successful aspects of the policy with regard to mountain ranges concerns the creation of institutions whose advice and proposals contribute to the development and protection of mountain regions. This is the case of the National Mountain Council and the Range Committees. The Prime Minister presides over the National Mountain Council which, through its permanent commission, coordinates public actions undertaken in mountain areas by the different ministerial departments. The range committees, of which there are seven (one per range), are presided over by the regional prefects; they define objectives and decide upon the actions that they consider necessary for the development of the ranges. The functions and make-up of these institutions have been defined by the 1985 Mountain Law and its implementing regulations.

In Viet Nam, a Committee for Ethnic Minorities and Mountainous Areas was created in 1993 in order to provide support to the mountain provinces and encourage the adoption of appropriate policies for them (FAO 2001). The committee collaborates in the implementation of the 1,715 Poor Communes Programme (1998–2005). The objective of this programme is to improve the precarious living conditions in the mountainous and remote areas. The methods planned for carrying this out include generating income and employment, improving infrastructures, and allocating administrative tasks at the local level (UNDP 2000).

Non-governmental structures

Governments are increasingly involving NGOs and associated organizations in the implementation of national policies and programmes. Because these organizations generally group together actors from different spheres of activity and are specialized in particular fields, they are often better able to meet local needs. They also provide an effective means of communication between the different interest groups, communities, and government authorities. They directly contribute in general to the formulation and application of national strategies and plans, as the following examples indicate:

• The actions undertaken by the Georgian Union of Mountain Activists clearly illustrate the type of interventions that an NGO can carry out in implementing a national, sustainable, mountain-development strategy. Having effectively participated in the preparation of the "Mountain Law" prior to its adoption in June 1999 by the Georgian Parliament, the Union plans also to actively contribute to the socio-economic development of the country's mountain regions. It also intends to provide full support to the application and perfecting of the Mountain Law.

- The Mountain Conservation Society of Tanzania (MCST) is very active in supporting the sustainable development of the country's mountain regions. Its activities are centred upon raising public awareness regarding the importance of safeguarding the cultural and natural heritage of mountain regions. The MCST, aware of the tourist potential of Tanzanian mountains, believes that the economic situation of mountain communities can be enhanced by establishing community-based ecotourism projects that make the most of the natural attractions and rich cultural heritage of these regions. Such projects have been successfully undertaken in the Uluguru Mountains since July 2000 (Mountain Conservation Society of Tanzania 2002). Tourists can benefit from the local culture and the daily mountain-communities' activities in the course of a special tour in typical mountain villages. Such ecotourism projects help to increase incomes of mountain communities.[3]
- In Switzerland, a country famous for its mountains, a number of different associations have played an important role for some time in mountain development. One of these is the Swiss Mountain Region Organization – an associative organization that has collaborated with public authorities since 1943. Its activities include the sustainable development of mountain resources, the protection of the mountain environment and its economic resources, and the furthering of equal treatment for both mountain and open country regions. The Swiss Aid to Mountain Peoples Association is another group that has supported the country's mountain regions since 1952. By concentrating its efforts on improving the economic potential and living conditions in mountain regions, this group seeks to halt the exodus of mountain peoples by supporting mountain agriculture and favouring mutual aid among mountain farmers. It also contributes to providing better housing and working conditions, improving the training of young farmers, supporting social assistance to people in difficulty, and helping to develop mountain regions by building appropriate infrastructures.

National IYM committees

On the occasion of the International Year of Mountains (IYM), numerous countries decided to create a specific structure charged with coordinating all the activities undertaken in this area, generally in the form of an IYM national committee. By the end of 2002, 78 countries had established (or were planning to establish) an IYM national committee or similar mechanisms. Although the composition of these committees varied from one country to another, they normally grouped representatives of government, NGOs, research institutes, and (frequently) parts of the private sector in areas that operate in the mountains – such as agriculture, forestry, tourism, handicrafts, the environment, and national and regional development. These committees were initially created with a precise objective – namely, to guarantee the coordination and effectiveness of all activities carried out for the IYM. Once this first objective had been accomplished, and following the events celebrating the IYM year (2002), the next task would be to capitalize on the accomplishments of these committees for the future. As they include representatives of the different mountain-region sectors, they would be expected to continue to encourage and coordinate sustainable mountain development in accordance with the institutional modalities to be defined, based upon the different national contexts.

Institutions with particular authority regarding mountains

Most countries do not have specific public institutions charged with sustainable mountain development. Institutional responsibilities are frequently divided among different ministerial departments – in particular, those devoted to agriculture, rural development, environment, national and regional development, water resources, and tourism. Although this institutional option often lacks coordination with regard to policies and sectoral actions, it occasionally succeeds in creating an overall and integrated approach to sustainable mountain development by achieving a degree of collaboration between the different ministries concerned.

In Croatia, where no specific body has been created for the administration of mountains, a number of public institutions have become involved in making decisions regarding mountain regions, particularly the Ministry of Environmental Protection and Territorial Development and the Ministry of Agriculture and Forestry. The latter Ministry has implemented a financial-assistance programme for the development of small agricultural holdings, including those situated in mountain regions.[4]

In Mexico, different ministerial departments are involved in mountain issues and are currently collaborating in implementing a regional sustainable mountain-development project by providing technical and financial assistance. The ministerial departments involved in this project are the Ministry of Social Development; the Ministry of Agriculture, Animal Breeding and Rural Development; and the Ministry of the Environment, Natural Resources and Fisheries.

In the Philippines, policy orientations for the mountains are found in sectoral policies concerning forestry, agriculture, tourism, and rural development, for example. As a result, the administration of mountain areas has often been characterized by an overlap of the national line agencies' functions and responsibilities. In order to remedy this, a Management Council for the mountainous Northern Negros Forest Reserve was created in 1996 as a policy-making and coordinating body, to protect, develop, and preserve the Reserve. The Council includes among its members representatives from the government, the civil society, academia, NGOs, and mountain communities. Although the Council focuses on social forestry, its programme has significant impacts on mountains: it promotes, among others, the implementation of an educational and informative campaign "to enhance the environmental awareness of the mountain population that will lead toward a community-based protection and conservation of the area" (Sánchez 2002).

Conclusions and recommendations

A few final suggestions can be drawn from the above observations and in the light of more general analyses on policy and institutional aspects of sustainable mountain development (e.g. Byers 1995; FAO 2000; Gabelnick et al. 1997; Lynch and Maggio 2000).

Policies and institutions at the service of sustainable mountain development

In order to establish sustainable mountain development on a solid basis, governments must create policy frameworks and institutional mechanisms that are tailored to the realities and constraints of the national context and meet the conditions and particular needs of mountain populations and ecosystems. Policy and institutional tools for mountain development must be created, implemented, evaluated, and carried out in a concerted manner, with the participation of all the actors – public and private – at the national and local levels who are involved and concerned.

Choosing realistic and optimal policy and institutional approaches

Governments must provide the appropriate means, both human and financial, in order to achieve the goals that have been set, whether by creating specific policies and institutions that are suitable for mountain regions or by incorporating mountain issues in their existing policies and institutions. The results that can be expected from the implementation of these policies will depend to a large degree on their ability to evolve and adapt to mountain conditions.

Placing mountain regions within a holistic-development perspective

The policies affecting mountains, whether sectoral (such as agriculture, forestry, water resources, animal breeding, tourism, mining) or transversal (such as environmental, or national and regional development) must be created and applied in a holistic-development perspective. Mountains must constitute a distinct entity and be recognized as such within this overall view. Inasmuch as approaches to natural-resources management have a direct impact on mountain ecosystems, this holistic approach requires the development needs of mountain regions and the sustainable management of natural resources to be taken into account.

Promoting an integrated approach to mountain development

In order to ensure coordination and harmonization, mountain policies must be developed and carried out in an integrated manner and in concert with other national policies regarding development and the environment. This integrated approach is doubly advantageous: it reduces the risk of conflict between the different policies dealing with the mountains and allows them to benefit from the financial contributions that could be mobilized through the integrated implementation of different policies.

Policies and institutions that attempt to meet the needs of local populations

The socio-economic needs of local communities and the concerns regarding the environment and sustainability in mountain regions can be combined by using a participative and decentralized "mountain approach." An approach of this nature would also make it possible to involve mountain people, NGOs, and local authorities in the decision-making process regarding mountain-region development.

Preserving mountain ecosystems and the cultural identity of their population

The importance and fragility of mountain ecosystems are among the principal concerns expressed at the international level and have become increasingly recognized as such by governments. This political recognition must result in the adoption of appropriate conservation measures. National policies must include precise measures favouring the preservation of mountain ecosystems and the safeguarding of the cultural identity of the communities that live there.

Enhancing indigenous knowledge and traditional techniques

Mountain communities, depositories of indigenous knowledge, have an intimate familiarity with their environment and possess a thorough awareness of its characteristics and limits. This knowledge, and the experience deriving from it, can be extremely useful in managing the natural resources of mountain regions. It is, therefore, indispensable to take this traditional knowledge into account when creating and applying sustainable mountain-development strategies and programmes.

Recognizing the role of women in mountain communities

Gender inequality is very accentuated in mountain communities. Although women frequently play a major role in managing natural resources, their contributions often go unrecognized. National mountain-development policies must therefore specifically acknowledge the important contribution made by women and must guarantee their equal treatment with other citizens.

Mountain-community participation in decision-making processes

Local populations are rarely (or minimally) consulted when decisions are to be made, although this has often been recommended. Political decisions regarding the mountains have direct consequences on the living conditions and means of subsistence of the people living there. The participation of mountain communities in decision-making processes must therefore be reinforced, in order to take into account their needs and concerns and to fully respect their rights.

Improving the living conditions of mountain populations

It is well known that particular conditions in mountain regions (e.g. climate, altitude) have a direct influence on the living conditions and quality of life of their communities. These communities often fail to receive adequate social service benefits, owing to their relative isolation. Particular financial efforts must, therefore, be made for these communities in the areas of education, health care, transport, and infrastructures in order to improve their living conditions and to bring them as close as possible to those enjoyed by the rest of the country.

Creating basic scientific databases relating to mountains

Given the importance of scientific information for appropriate mountain management, and in view of data gaps that often hinder proper mountain-development planning, it may be advisable to create national databases with basic scientific mountain information. These may include, for example, data on agricultural infrastructures and socio-economic facilities to help planners and policy makers to formulate effective planning of agricultural activities in mountain areas.

Dissemination of the successful results in sustainable mountain development

Given the recently acquired awareness of the need to promote sustainable mountain development, it is vital to disseminate the successful results of policies, strategies, and national programmes in this area to all the decision makers and actors concerned. The Mountain Forum website [www.mtnforum.org/emaildiscuss/discuss02/042602443.htm] could be an excellent medium for exchanging pertinent information and experiences regarding sustainable mountain development.

Notes

1. United Nations Sustainable Development website: Japan Profile (for the Johannesburg Sustainable Development World Summit: http://www.un.org/esa/agenda21/natlinfo/wssd/japan.pdf).
2. The IYM Italian Committee website: http://www.montagna.org/montagnachevince/prioritam.asp
3. MCST case study for BGMS-A2: implementation of ecotourism projects in Uluguru mountains (http://www.mtnforum.org/emaildiscuss/discuss02/042602443.htm) and Geocities Website: www.geocities.com/mountainconservers/Tanzania_mountains.html
4. United Nations Sustainable Development Website, Croatia Profile (for the World Summit on Sustainable Development, Johannesburg 2002): http://www.un.org/esa/agenda21/natlinfo/wssd/croatia.pdf

REFERENCES

Byers, E. 1995. *Mountain Agenda: Environmentally sustainable and equitable development opportunities.* Franklin, West Virginia: The Mountain Institute. [http://www.mtnforum.og/resources/library/byere95a.htm]

Dax, T. 2002. *Endogenous development in Austria's mountain regions: From a source of irrigation to a mainstream movement.* Vienna. [www.mtnforum.org/resources/library/daxth02a.htm]

FAO. 1998. *Integrated coastal area management and agriculture, forestry and fisheries.* Rome: FAO.

FAO. 2000. *International Year of Mountains. Concept paper.* Rome: FAO.

FAO. n.d. Task Manager Report. *Agenda 21, Chapter 13, Sustainable Mountain Development, Part II: Policy Report.* Rome: FAO.

FAO. 2002. *Cross-sectoral linkages in mountain forest development – Italy case study.* Rome: FAO.

FAO and Italy Cooperative Programme. 2000. *Mountain people in sustainable development*. Rome: FAO.

Fatyga, J. 2002. *Policy of the Republic of Poland towards a sustainable development of the mountainous areas*. Case study BGMS-A2.

Friend, D.A. 2002. *Regulating development through water quality standards*. Telluride, Colorado (USA). Case Study BGMS-A2. [http://www.mtnforum.org/emaildiscuss/discuss02/050202474.htm]

Gabelnick, T. et al. 1997. *Managing fragile ecosystems: Sustainable mountain development. Policy analysis and recommendations*. Princeton University, USA. [http://www.mtnforum.org/resources/library/gabex97a.htm]

Giurova, E. 1995. *Régions de montagne bulgares: Les enjeux de leur intégration dans la coopération européenne*. Extraits des Actes de la Conférence de Cracovie. [http://www.euromontana.org.Bulgarie.htm]

Joshi, A.L. 1997. *Community forestry in Nepal: 1978 to 2010*. Nepal: Ministry of Forests and Soils Conservation. [http://www.mtnforum.org/resources/library/josha97a.htm]

Lynch, O.J. and G.F. Maggio. 2000. *Mountain laws and peoples: Moving towards sustainable development and recognition of community-based property rights*. Center for International Environmental Law, Washington, D.C.

Meknassi, R.F. 2000. *Etude sur la loi montagne. Rapport préliminaire: Version provisoire*. Rabat (Maroc).

Ministère chargé des eaux et forêts (Maroc). 2000. *Politique pour la protection et le développement de la montagne*. Rabat.

Ministère de l'aménagement du territoire et de l'environnement (France). 2001. *Les grands programmes du développement durable*. [www.environnement.gouv.fr/ministere/rapportactivite/2000/Datar/prog-dev-durable.htm#hautpage.]

Mountain Conservation Society of Tanzania. 2002. *Implementation of ecotourism projects in Uluguru Mountains – Tanzania*. Tanzania: MCST. [http://www.mtnforum.org/emaildiscuss/discuss02/052202557.htm]

Perera, F.R. 2002. *"Síntesis de los 15 años de trabajo del programa de desarrollo integral y sostenible de las montañas cubanas."* Cuba. [Unpublished.]

Pradhan, P.K. 2002. *Spatial policy issues in agriculture development in the mountains of Nepal*. Nepal. [www.mtnforum.org/resources/library/pradp02a.htm]

Rivera, B. (Grupo ASPA). 2002. *Politicas de conservación de áreas protegidas*. Colombia. [English version available on the Mountain Forum website: http://www.mtnforum.org/emaildiscuss/discuss02/052102551.htm]

Sánchez, B.Q. 2002. *The Northern Negros Forest Reserve Management Council: A multi-stakeholder, multi-functional approach*. Philippines.

UNDP. 2000. *Compendium of Rural Development Assistance in Viet Nam*. Ha Noi.

Villeneuve, A., A. Castelein, and M.A. Mekouar. 2002. *Mountains and the Law: Emerging trends*. Rome: FAO.

10

Prospective international agreements for mountain regions

Wolfgang E. Burhenne

Summary

There are no international agreements that specifically address mountain concerns (especially with a view to achieving sustainable development), except for the Alpine Convention. This can be taken only as an example, not as a model, for other specific mountain-ecosystem agreements.

In considering the development of new international legal instruments for mountain regions, the numerous general international environmental agreements relevant to mountain regions that are currently in force have to be taken into account; therefore, one should take care not to create rights and obligations that conflict with these general rules. Several non-legally binding instruments also offer guidance for mountain regions.

In designing any further accord for a particular region, it is necessary to evaluate the natural conditions as well as the legal, political, economic, social, and cultural situations; to identify the specificities of the mountain ecosystem; and to consider which aspects require a transboundary rule or harmonized conduct, while taking the principle of subsidiarity into account. During the drafting process, special attention should also be paid to ensure that the form is suitable for a legally binding instrument and is not merely of a declaratory nature.

A threshold discussion and decision should also be devoted to the format of the prospective agreement – whether it is to be a comprehensive or a framework instrument, that sets out general principles to be followed by additional legal instruments or decisions by the parties targeted at specific subject areas.

A check-list for possible elements for regional mountain-ecosystem agreements is provided. This list cannot be exhaustive and is not intended to be prescriptive: such a list simply cannot take into consideration all the natural, socio-economic, and political conditions that are relevant to a specific situation.

Introduction

In considering the development of new international legal instruments for mountain regions, a number of factors have to be taken into consideration.

No legally binding global agreement specifically covers concerns related to mountains in general. This is evidence of the fact that the natural conditions – as well as the legal, political, economic, social, and cultural situations – vary considerably for individual mountain regions. However, numerous global and regional agreements currently in force address the conservation of ecosystems, the sustainable use of natural resources, and the processes and activities that affect both, with the general and overarching aim of achieving sustainable development. These multilateral general environment agreements (hereafter referred to as general MEAs) are naturally also relevant to mountain regions, in so far as they are in force for the states in the region considered (Fodella and Pineschi 2002; Lynch and Maggio 2000).

In designing any further accord specific to any particular mountain region, these general MEAs should therefore be taken into account. In particular, international agreements for any mountain region should not duplicate rules already agreed in general MEAs. Most importantly, these mountain agreements should not create rights and obligations that conflict with general MEAs. However, agreements designed for mountain regions can install measures to support MEA implementation in a specific context and can introduce rules that are complementary to (and are more far-reaching than) the MEA in question, in a way attuned to the situation and needs of the region considered. Guidance for doing so is provided in a global, but non-legally binding, instrument – Chapter 13 of Agenda 21 entitled "Managing Fragile Ecosystems – Sustainable Mountain Development," adopted at the United Nations Conference on Environment and Development in 1992 (United Nations 1992). These recommendations are useful guidelines in drafting agreements for any mountain region and, for this reason, are taken into account in the check-list presented below.

It derives from the above that any agreements for mountain areas should focus on aspects upon which it is important to agree at the level of the mountain region considered – i.e. for which a limitation of national sovereignty is needed to effect
• a transboundary rule, or
• a harmonized conduct
in order to achieve sustainable development of the region as such. The principle of subsidiarity should be recognized with a view to avoiding overregulation and thus easing negotiations: international rules are necessary only if the parties to the negotiations agree that they are desirable in order to achieve a common goal.

A threshold discussion and decision concerns the format of the prospective agreement. There are two fundamental options:
1. a comprehensive instrument that covers all obligations in the areas dealt with in the treaty considered;
2. a framework treaty setting out general principles and ground rules, to be elaborated in "additional" protocols addressing specific subject areas.

A comprehensive instrument, once adopted, permits a more rapid start on implementation on all fronts; however, it takes a longer negotiation period, as well as a wider array of expertise during these negotiations. Failure to agree on all issues at once endangers the conclusion of the entire agreement. A framework instrument is easier to conclude, as it concentrates on setting up the scope and basic requirements of each of its elements, leaving detailed rules for future protocols. Although negotiations of the protocols may take a considerable time, implementation of the framework core may start, pending their conclusion. Furthermore, because protocols are more specific and detailed, they need not be modified often. Experience tends to indicate that the framework approach is a more practical way of achieving a set of goals, step by step; however, in order to do this the goals must be clearly identified in the framework.

In Europe, all the above-mentioned points have been considered during the negotiation of the Convention for the Protection of the Alps (Alpine Convention), signed in 1991. It can, therefore, be taken as a good example of a specific mountain-ecosystem agreement and a source of inspiration for agreements elsewhere. The Alpine Convention is a framework convention which, in Article 2, requests the conclusion of Protocols with concrete formulated targets. Most of these protocols have been concluded; a few are still in preparation (Kiss 2002; see www.cipra.org). Although the specificity of each mountain area makes it unrealistic to take the Alpine Convention as a model for other possible mountain-related agreements, the experience gained in its development has been used in the development of other regional instruments, such as the Carpathian Convention signed in 2003 (see Angelini, Egerer, and Tommasini 2002) and others under development, for instance for the Altai and Caucasus. The text of many relevant instruments may be found in Treves, Pineschi, and Fodella (2002).

In drafting any further regional agreements on mountain ecosystems, a check-list of items that need to be considered and evaluated may be useful. In particular, such a list may provide a basis to decide whether, in the particular context, regional norms are needed and should be negotiated accordingly. During the drafting process, it is important to check constantly whether a formulation under proposal is to be regarded as legally binding or is merely of a declaratory nature (which is not appropriate for an agreement, except for the preambular paragraphs). In addition, it may be possible to forgo negotiating legally binding text in instances where a consensus can be reached on procedures that concern only the contracting parties. For example, in the case of the Alpine Convention, it was agreed that there is no need for a special protocol on compliance procedures.

Such a list is provided below. A number of provisos are in order:
- Such a list cannot be exhaustive, because it simply cannot take into consideration all the natural conditions that may be applicable to a specific situation;
- The content and the structure of the document, as well as the elements listed below, are by no means prescriptive, as it is ultimately up to the prospective parties to decide which points are relevant to their concerns, and in which order to tackle them.

Possible elements of a regional mountain ecosystem agreement

A. Preamble

- Preambular clauses depend on the type and content of the agreement under consideration.

B. Objective(s)

- Achievement of integrating of the environmental, economic, and social dimensions;
- Development of a coherent regional policy;
- Development of economically weak areas;
- Harmonization of requirements (e.g. permit criteria).

C. Fundamental principles

- Common concern;
- Intergenerational equity;
- Cooperation;
- Sustainable development;
- Prevention;
- Precaution;
- Information and participation of the public and communities.

D. General obligations

- Protection, management and restoration of nature and landscapes;
- Promotion of sustainable livelihoods, *inter alia* through gender-sensitive policies;
- Maintenance of the identity of various cultural groups;
- Application of harmonized obligations throughout the treaty area;
- Development of mountain-specific quality goals;
- Duty to consult in case of potential transboundary interference.

E. Specific commitments/obligations

Planning instruments and mechanisms

- Assessment mechanisms, e.g. inventories and maps;
  - Identification of fragile, sensitive, and special-problem areas.
- National and regional strategies, plans, and programmes;
  - Careful allocation of uses;
  - Transboundary coordination.
- Land-use planning;
  - Land-use planning mechanisms for the integration of social, economic, and environmental aspects;
  - Binding zoning plans (designating areas for specific uses – e.g. agricultural, residential, and tourist while planning for biological diversity conservation);
  - Mitigation measures for unavoidable impairments;
  - Limitation of secondary residences;
  - Construction/building guidelines;
  - Creating financial mechanisms, for example setting aside a portion of the proceeds from profitable ventures (such as projects that add value to land) in order to create funds for ecoservices, and disaster relief and/or prevention;
  - Rights and obligations of local communities;
  - Rules regarding property rights.

Other tools

- Promotion of traditional mountain economies, small-scale production;
- Use of tax and market-economy incentives/disincentives to improve management and sustainable use of natural resources;
- Use of existing indicators for sustainable development (see UN Department for Statistics), as well as development of new ones for mountain-specific environmental-quality goals;
- Procedural rights;
- Environmental impact assessment (EIA), risk assessment;
- Promotion of role of non-state actors in implementation;
- Marketing of local products (quality labels/labels of origin);
- Establishment and maintenance of transboundary sectoral networks (e.g. natural-disaster relief).

Managing natural and cultural resources

- Soils:
  — Maintaining multifunctionality of soils;
  — Ensuring secure land tenure and access to land, credit, and training;
  — Removing obstacles that inhibit farmers (especially small-scale farmers) from investing in and improving their lands and farms;
  — Combating soil degradation (e.g. erosion):
    - grazing restrictions
    - keeping livestock at appropriate levels and in appropriate areas (carrying capacity) taking into account the needs of wild species
  — Prevention of soil pollution (including rehabilitation of contaminated sites);
  — Restoration measures.
- Forests:
  — Conservation of natural forests;
  — Sustainable use of forests:
    - access to and use of environmentally sound technologies
    - regulated grazing
    - timber concessions
    - ecologically sound afforestation and reforestation
    - promotion of natural regeneration
  — Conservation of forest for ecological and other functions (prevention of landslides and avalanches);
  — Combating deforestation and forest degradation;
  — Rehabilitation and conservation strategies for countries with low forest cover;
  — Rehabilitation and restoration of degraded forest lands;
  — Transboundary cooperation in case of forest fires.

- Water resources:
  — Protection of the availability and quality of water resources:
    - groundwater
    - wild streams
    - rehabilitation of streams
    - drinking-water reservoirs
    - wetlands conservation
  — Water for sustainable food production and sustainable rural development;
  — Integrated Watershed Development Programmes:
    - dams and redirection of watercourses
    - enhance effective participation of local communities
    - ensure that downstream communities benefit from upstream activities.
- Biological diversity and landscapes:
  — Biological diversity:
    - wild species conservation and sustainable use:
      • monitoring populations and habitats
      • species management and habitat conservation:
        • harmonization and coordination of protected-species measures
        • transboundary management plans
        • sustainable management of ungulate population
      • controlling processes and activities with potential negative impact on species (e.g. the introduction of alien species)
    - domestic species
  — Genetic resources and related traditional knowledge;
  — Creation/maintenance and management of protected areas:
    - harmonization of protected-areas types
    - transboundary protected areas
    - networking between protected areas
  — Landscapes:
    - maintenance of cultural and natural landscapes
    - landscape planning (see also land-use planning)
      • green areas/spaces
    - restoration and management
  — Management tools for biological diversity and landscape conservation:
    - ecosystem-based management
    - community-based management.
- Cultural heritage:
  — Maintenance of diversity of traditional/indigenous cultures (languages and customary ways of life);
  — Preservation of cultural sites and structures/buildings.

Managing processes and activities

- Rural development:
  — Best practices in land-resources management to achieve sustainable food cycles;
  — Improvement of working conditions in mountain areas;
  — Fostering rural development with the emphasis on, *inter alia*, socio-economic diversification, employment, capacity building, participation, poverty eradication, empowerment, and partnerships;
    – preserving environmentally sound traditional methods as well as promoting new ones
    – choices in agricultural production techniques, consumption patterns, and safety regulations
    – knowledge for a sustainable food system: identifying and providing for education, training, knowledge sharing, and information needs
    – eliminating perverse incentives.
- Tourism:
  — Promotion of environmentally friendly tourism infrastructures;
  — Limiting leisure activities with potential negative impact on the immediate environment (such as mountain biking, paragliding, ice-climbing, heli-skiing, artificial snow machines) to specific areas while imposing strict bans elsewhere, especially in protected areas;
  — Participation of local and indigenous communities in decision making with a view to preventing activities interfering with traditional way of life or affecting their livelihood;
  — Cooperation with countries from which tourists originate;
  — Influence streams of tourists (e.g. through diversifying school vacation periods).
- Non-renewable resources:
  — Extraction;
  — Rational use.
- Energy:
  — Ensuring adequate energy infrastructure for local populations;
  — Decentralization of energy supply sources;
  — Energy-conservation measures;
  — Capacity-building and technology transfer to promote renewable energy sources;
  — Ensuring efficient use of fossil fuels.
- Transport (infrastructure):
  — Promotion of efficient public transportation;
  — Transboundary cooperation on regulating transit traffic;
  — Tolls for foreign cars and/or heavy loads;

— Promotion of alternative modes of transportation that are more eco-friendly.
- Pollution:
  — Environmentally sound management of domestic and industrial waste;
  — Hazardous substances and wastes;
  — Prevention costs (e.g. "polluter pays" principle);
  — Restoration and compensation measures;
  — Promotion of less-polluting substances;
  — Introduction of topographically related regulations for noise pollution.
- Disaster reduction:
  — Prevention of adverse impact of natural hazards and of disasters caused by human activity:
    – good planning, including land-use planning
    – education and training of stakeholders
    – early-warning systems and forecasting:
      · making latest technology available
    – effective evacuation infrastructures
    – Transboundary Cooperation for Disaster Relief:
      · flight-space permits and access of specialist teams to neighbouring territories
      · setting up emergency funds
  — Mitigation through measures to limit adverse impact:
    – Restoration costs and/or compensation for environmental damage:
      · Trust Fund for EcoServices (see above)
      · Additional funds set aside from community revenue.

F. Implementation mechanisms

- Participatory mechanisms (e.g. through existing subregional institutions);
- Designation of national authorities/focal points responsible for implementation;
- Compliance procedure;
- Dispute settlement.

G. Institutional and financial aspects

- Determining the institutional machinery and its functions:
  — Conference of the Parties (COP);
  — Subsidiary bodies;
  — Secretariat.

- Possibility of rotating organizational and financial responsibilities between parties (based on a rotating Presidency for a specific period):
  — Participation of NGOs.
- Determining COP rules and procedures:
  — In particular, budgetary rules (e.g. making regular budget share contingent to each party's share of total mountainous territory, its share of total population inhabiting the area and/or total GDP).

H. Final clauses

- Signature, ratification, and accession;
- Entry into force;
- Amendments;
- Depositors.

REFERENCES

Angelini, P., H. Egerer, and D. Tommasini (eds). 2002. *Sharing the Experience: Mountain Sustainable Development in the Carpathians and the Alps.* Bozen: Europäische Akademie Bozen.

Fodella, A., and L. Pineschi. 2002. "Environment Protection and Sustainable Development of Mountain Areas." In: T. Treves, L. Pineschi, and A. Fodella (eds) *International Law and Protection of Mountain Areas.* Milan: Giuffrè Editore.

Kiss, A.C. "Place et role du système conventionnel alpin dans le developpement du droit international de l'environnement." In: T. Treves, L. Pineschi, and A. Fodella (eds) *International Law and Protection of Mountain Areas.* Milan: Giuffrè Editore.

Lynch, Owen, and Gregory F. Maggio. 2000. *Mountain laws and peoples: Moving towards sustainable development and recognition of community-based property rights.* A general overview of mountain laws and policies with insights from the Mountain Forum's e-conference on Mountain Policy and Law. Franklin, West Virginia: The Mountain Institute, Center for International Environmental Law, and Mountain Forum.

Treves, T., L. Pineschi, and A. Fodella (eds). 2002. *International Law and Protection of Mountain Areas.* Milan: Giuffrè Editore.

United Nations. 1992. *Agenda 21: Chapter 13: Managing Fragile Ecosystems: Sustainable Mountain Development.* UN Doc. A/CONF.151/26 (Vol. II), 13 August 1992.

11

The role of culture, education, and science for sustainable mountain development

Bruno Messerli and Edwin Bernbaum

Summary

Culture has a key role to play in sustainable mountain development. People feel deeply motivated to conserve natural resources valued by their religious and cultural traditions. Over the centuries, many cultures have developed beliefs and practices that sustain mountain environments. These cultures have an intimate knowledge of mountain resources that needs study and recognition as intangible heritage.

Modernization, population growth, and globalization threaten to overwhelm traditional beliefs and practices that have sustained mountain environments. Local communities need to find ways to strengthen these beliefs and practices and to adapt them to changing circumstances. Traditional elders and healers should be included in educational and scientific programmes.

Tourism managed in culturally appropriate ways can help to sustain mountain cultures economically. Programmes that support arts and crafts can also benefit local communities. Another source of revenue is the knowledge of medicinal plants. In the development of these and other mountain resources, local communities need to receive an equitable share of the benefits.

Like living organisms, cultures evolve and change. They can accommodate to shifting circumstances and contribute to programmes of mountain development in sustainable ways, but only if they maintain their integrity.

Education for environment and development is an important topic in Agenda 21. Although most countries address this topic in school curricula, environmental-education programmes need to be further strengthened, especially in mountain areas. Teachers without a good knowledge of the cultural and environmental conditions, but also without the necessary practical and didactic qualifications, will never be able to make the village school a creative centre of sustainable development for the young or even for the older generation.

A college education should be open for well-qualified children, with equal opportunities for girls and boys. The main aim should not be to increase the number of jobless graduates but to prepare individuals for leading positions in all fields of mountain development. Unless educational opportunities are linked with avenues for personal advancement, it is difficult to retain young people in remote mountain areas. University education must create a new responsibility for nature and society and for improving basic needs in poor and remote mountain areas of the developing world, while capacity-development programmes, especially in developing countries, are also vital for the sustainable development of mountain resources.

Modern information technologies, special information libraries, and multimedia centres are promising initiatives to overcome the digital and scientific divide between the rich and the poor parts of the world. As a consequence, new methods and tools such as distance learning and open universities must be seriously taken into consideration, though this may require some experimentation.

With regard to science, until the 1970s mountains were considered as marginal by the leading natural and social sciences. This attitude has changed with the rapidly growing interest in environmental problems, natural resources, and mountain societies. The close relationships between natural processes and human activities in mountain areas has created a high demand for interdisciplinary and transdisciplinary research, including traditional knowledge. Attempts have been made to define the research priorities for the future, in particular regarding water resources, biological diversity, land degradation, economic standards, mountain cultures, and tourism, as well as cultural and climate changes.

For the future, we shall need scientists who understand, besides their specialty, how both natural and human systems operate and interact – scientists who can think locally, nationally, and globally. The International Year of Mountains must be the beginning of a new research effort towards sustainability science for mountain environment

and development and for the highly complex highland–lowland interactions.

Introduction

Mountain areas are characterized by close interactions between natural
processes and human activities, and by a sometimes difficult relationship
between highlands and lowlands. In the last twenty years, and even more
so since the Rio Earth Summit in 1992, new driving forces have become
evident: these are globalization, urbanization, a growing divide between
the rich and the poor parts of the world, from the human side; and
climate–environmental change from the natural side. Traditional knowledge, the experience of generations over centuries, has been overrun by
these external driving forces reinforced by modern information and
communication technologies. The consequences for mountain ecosystems
and mountain communities are not yet known. Understanding these processes, and managing this growing complexity for sustainable development, requires an intimate knowledge of the cultural conditions, new
ideas and tools for education, and new objectives and approaches for
science.

Culture

Background and issues

From the Andes to the Himalaya, mountain cultures around the world
are inextricably intertwined with the landscape, each influencing and
shaping the other in complex ways. Programmes of sustainable mountain
development need to recognize that people are an integral part of the
environment and that their diverse ways of life need to be sustained and
developed along with biodiversity and natural resources.

Culture has a key role to play in sustainable mountain development.
Over the centuries, many traditional cultures have developed beliefs and
practices that preserve mountain environments. These beliefs and practices often provide a sounder, more enduring basis for conservation and
development than measures based solely on economic, legal, or scientific
considerations.

Programmes of sustainable mountain development need to take cultural values, traditions, and preferences into account: if they do not, they
will fail to engage local communities and other stakeholders whose support they need to be truly sustainable over the long term. Poorly de-

signed conservation measures and development policies that undermine mountain cultures can make matters worse by weakening traditional practices and controls that have allocated resources and protected the environment for generations.

Women play particularly important roles in passing on and maintaining mountain cultures through family life. In many mountain cultures, they have roles that give them a particular stake in sustainable-development issues. In some Himalayan regions of Nepal and India, for example, they are the ones who have to travel further and further from their homes to gather fodder and firewood: as a consequence, they take a more active interest than do men in restoring and protecting dwindling forests (Messerli and Ives 1997).

In a number of mountain societies, such as the Sherpa and Tibetan, women have greater equality than their counterparts down on the plains. In other cultures, such as the Pashtun in Afghanistan, women are severely repressed and, under the Taliban, they have not been allowed to work or to go to school. Supporting traditional cultures in such cases can conflict with promoting women's rights and equality, especially when gender distinctions are central to a culture's core values.

Knowledge

The relative isolation of many mountain communities has allowed them to retain traditional cultures and ways of life abandoned in the more accessible lowlands. Descendants of British settlers in the Appalachian Mountains have preserved old variants of ballads no longer found in Britain. Dissenting groups with cultures opposed to mainstream societies, such as the Albigensians in Europe, have sought refuge in mountains such as the Alps in order to preserve their cultural integrity and to practise their religious beliefs. The promise of sanctuary in legendary hidden valleys has moved Tibetans to settle the Himalayan border regions of Tibet (Bernbaum 1998).

Because of their vertical topography, mountains have extremely diverse environments and microclimates. This diversity of terrain and climate provides niches for a great variety of mountain cultures based on different ways of life, many of them in close proximity to each other. Nomads grazing herds on the Tibetan Plateau, for example, exchange milk and meat for barley grown by farmers in more sheltered valleys. The natural role of mountain ranges as borders between nations and cultures adds to the rich cultural diversity of mountainous regions. It contributes, in particular, to the great number of languages and dialects found in mountains (Posey 1999).

Mountain cultures have used their traditional knowledge and practices to protect the environment in a variety of ways. The Dai people of south-western China have set aside their Holy Hills as gardens of the gods off-limits to hunting and farming, making them sanctuaries of biodiversity (Shengji 1993). When Dineh (Navajo) singers or healers collect medicinal plants, they pick them in rituals that minimize damage to the ecosystem (Posey 1999). Because wildlife belongs to the apus or mountain deities of the Peruvian Andes, many indigenous people of the region refrain from hunting (Bernbaum 1998). Traditional designations of sacred "lama for-ests" and the local institution of village forest wardens have done a better job of protecting forests in the Khumbu region near Mount Everest than have more recent measures instituted by the central government and Sa-garmatha National Park (Stevens 1993).

Cultural values in modern societies have also contributed to mountain protection. Mountainous national parks in North America, such as Yo-semite and Mount Rainier in the United States, were established, in part, as places for people to visit for spiritual and physical renewal. Public outrage throughout Europe forced the cancellation of a project to build a resort and cable car on Mount Olympus: Nobel laureates joined others in writing letters to the Greek Minister of Culture, protesting against the desecration of a symbol of Western culture (Messerli and Ives 1997).

Implications and recommendations

The advent of modern communications has opened even the most isolated communities to the outside world. Modernization, population growth, and globalization are additional factors tending to undermine many mountain cultures and to overwhelm traditional beliefs and practi-ces that have protected mountain environments. Local communities need to find ways to strengthen these traditional beliefs and practices and to adapt them to changing circumstances and outside influences. Govern-ment agencies, NGOs, and other organizations have a role to play in this process.

As a first step, any programme of sustainable mountain development should include representatives of local communities and other groups for whom the site under consideration has cultural significance. These stake-holders need to be involved from the beginning as full participants in the process. Their needs and priorities should take precedence in project planning and implementation (The Mountain Institute 1998).

NGOs and government agencies can help by recognizing the knowl-edge and authority of traditional leaders and experts charged with main-taining cultural traditions and protecting the environment. At hospitals on the Navajo reservation in the United States, for example, Western doctors practise side by side with traditional healers, drawing on both scientific medicine and traditional healing rituals for the benefit of their patients. Since these rituals lie at the heart of Navajo religion, this also helps to reinforce Navajo culture.

Programmes of sustainable development need to sustain mountain cultures along with natural resources and the environment. A natural source of income for many mountain communities is tourism, but it needs to be managed in culturally appropriate ways. Ceremonies that have deep meaning for local people can easily be reduced to superficial shows for the benefit of visitors. Mass tourism can have negative impacts: huge numbers of tourists in noisy buses have made the practice of contemplation impossible at the spectacular monasteries of Meteora in Greece (Messerli and Ives 1997).

If tourism and pilgrimage are to benefit mountain communities over the long term they must respect and enhance the integrity of their cultures and environments. Measures need to be developed to deal with the numerous pilgrims and tourists who have desecrated major pilgrimage sites – such as the Badrinath in the Indian Himalaya and the sacred mountain of Tai Shan in China – with deforestation, litter, and sewage. Negative cultural and environmental impacts will eventually destroy what makes such sites attractive to tourists and pilgrims in the first place (UNESCO 2001).

Arts and crafts express beliefs and practices underlying many mountain cultures. Programmes that support them can benefit local communities both economically and culturally. NGOs, government agencies, and private companies have a key role to play in providing access to outside markets for these traditional products. They can also help to ensure that locally produced arts and crafts are made available to tourists and pilgrims.

Another source of revenue that can help mountain communities sustain their cultures is the knowledge and collection of medicinal plants; however, if this is done for purely commercial reasons, without regard for traditional restraints, it can have adverse environmental and cultural impacts. Science can help: a Himalayan research station of the High Altitude Plant Physiology Research Centre in India is isolating efficacious strains of medicinal plants and determining ways to cultivate them at lower altitudes in order to reduce pressures on fragile alpine meadows (Körner and Spehn 2002).

Biodiversity and indigenous knowledge in the mountains of Tanzania

North-eastern Tanzania's Usambara Mountains have been a focal point of recent efforts aimed at the sustainable development of the region's tropical moist forests.

These are an important component in the multiple livelihood strategies of local people and also fulfil an important function in providing a stable water resource to densely populated downstream coastal areas, including Tanzania's second-largest mainland port, Tanga town. Planning is under way to identify and strengthen the benefits of the watershed's forests to local stakeholders. Non-timber forest products (NTFPs) have met, and continue to meet, local villagers' needs for energy, construction, food, and health. Sustainable NTFP use is considered to be an incentive that directly links the conservation of the mountains' forests to the livelihood strategies of the mountain villagers. Indigenous knowledge of medicinal plants and their application have a unique role in this planning.

In an innovative response to the region's human immunodeficiency virus/acquired immunodeficiency syndrome (HIV/AIDS) crisis, traditional healers and Western medicine have pooled their resources in a programme that treats HIV/AIDS patients with plant remedies. Since 1990, this collaborative effort has been operating under the auspices of the Tanga AIDS Working Group (TAWG), which currently provides care to 400 HIV/AIDS patients. These remedies, which improve condition and reduce the incidence of opportunistic infections, are made available to patients preferring traditional medicine.

In Tanzania, pharmaceuticals remain unaffordable for many people. Many HIV/AIDS patients are cared for at home, as traditional healers far outnumber medical doctors. An estimated 15 per cent of adults in Tanzania are HIV positive; in the Tanga region, which includes the Usambara Mountains, the proportion is possibly much higher. TAWG is a practical response to the epidemic, based on indigenous knowledge of the region's ecosystems. As a result, the rapidly growing epidemic is generating a high value for medicinal plants.

In this case, traditional knowledge is recognized for its important value, not only providing a direct contribution to public health but also justifying the need to maintain the culture and the ecosystems that are often deeply integrated. By meeting the health needs of HIV/AIDS victims, indigenous knowledge of medicinal plants adds value to the ecosystems in which they are produced, creating opportunity and incentive for mountain villagers to sustainably manage these productive resources. This exemplifies the positive potential for linkages, as the growing demand for medicinal plants is simultaneously leading to efforts to support NTFP production.

TAWG and other organizations have recognized the critical role of traditional healers where other forms of medical response are not available. With decreasing natural stocks of medicinal plants, traditional healers are now interested in the sustainable management of these important NTFPs and the conservation of the region's biodiversity. However, the indigenous knowledge of these plants and how they may be used in healing applications is threatened, as many with this important knowledge may die before passing it on to younger generations. Source: Marc Barany and A. L. Hammett, Virginia Tech. Contribution to online conference: http://www.mntforum.org/bgms/paperd2.htm

In the development of these and other mountain resources – such as forests, minerals, food crops, and water – local communities need to benefit economically in order to maintain the material bases for their cultures. Government agencies, NGOs, and private corporations should help them get a fair share of the revenues generated from their products. Local communities should also be compensated for their cultural knowledge and expertise, whether this takes the form of return on intellectual property – a modern legal concept – or something more in keeping with their own ideas of traditional ownership (Kemf 1993).

Programmes of sustainable mountain development can draw on traditional and modern cultures to ground their conservation efforts in deeply held values and beliefs that will make them more understandable and enduring. An innovative programme at the pilgrimage shrine of Badrinath in India has scientists from the G. B. Pant Institute of Himalayan Environment and Development working with priests to motivate pilgrims to plant trees for reasons that come out of their religious and cultural traditions (Bernbaum 1997). The Mountain Institute is working with the US National Park Service to develop interpretative and educational materials that encourage conservation based on the cultural and spiritual significance of different features of mountain landscapes and ecosystems in American, Native American, and other cultures around the world (Bernbaum 2000).

We need longitudinal studies on the long-term effects of various measures and programmes that make use of culture and religion in sustainable mountain development. For example, do local people and communities take better care of seedlings planted in religious ceremonies and refrain from cutting them down when the trees are fully grown?

Conclusions

Like living organisms, cultures evolve and change, adapting to shifting environments and circumstances. They can accommodate programmes of mountain development and contribute to them, but they can do so in a sustainable way only if they uphold the underlying principles that support and sustain their integrity.

Cultural diversity needs to be maintained as an intangible heritage and for what it offers to the rest of the world. Mountain cultures have much to contribute in terms of their knowledge and their ways of living in harmony with nature. They remind us that there are many ways of seeing the environment and many reasons for valuing and protecting the world in which we live.

Education

As stated in Chapter 36 of Agenda 21, "Basic education is the underpinning for environment and development education. All countries should strive for universal access to education, and achieve primary education for at least 80% of all girls and boys through formal schooling or non-formal education. Adult illiteracy should be cut to at least 1990 level, and literacy level of women brought into line with those of men" (Keating 1993). Five years after Rio, during a UNESCO conference about education in Thessaloniki, the following statement was addressed: "Education – a forgotten priority of Rio?" Even ten years after Rio, education for sustainability has not yet been adequately established, neither in the rich nor in the poor parts of the world. Moreover, there is frequently a hiatus between what is taught in schools and colleges and the actual economic policies of governments concerning the environment. Political decision makers should be the first to receive education for sustainability.

The following thoughts focus predominantly on mountain regions in the developing world, where inequalities between urban centres of the lowlands and remote mountain areas are extremely great. In these situations, it must be very clearly said that every improvement of education and school systems can lead to emigration on the local level and to "brain-drain" on a national level. There is no choice in our rapidly changing world but to keep these negative consequences in mind while improving the education system for sustainable development for people of all ages.

Primary-school education depends on the education of the teachers

The first question is whether a national education programme for remote mountain areas exists, and whether it is in a position to maintain village schools in remote mountain areas. The second question is whether there are teachers willing to spend several years in such places, which are accessible only with difficulty and therefore may affect the teachers' family lives. One solution would be to train local inhabitants who are more likely to stay on, instead of bringing in teachers from outside. More flexibility in the formal educational qualifications is needed, but this is likely to be compensated for by the greater commitment that comes from belonging to the region and understanding its culture and environment. Only if these conditions are fulfilled can the third question be raised: whether the education of teachers is sufficient to educate the next generation in such a harsh environment. Research reports about teachers' aims and methods indicate that, in primary schools, teacher-led, descriptive, and theoretical teaching is more common than learner-centred, enquiry-based, active participatory teaching connected to problems of the surrounding human conditions (Forsyth 2001). Answering these three questions and improving education in mountain areas with positive effects for the whole community is a national responsibility; better education of the teachers is a university responsibility. Foreign aid to improve the basic education system in a developing country should be the last resort if there are absolutely no national resources available, as dependence on foreign aid has often been an obstacle to educational development, independence, and sustainability in less-developed countries. Nevertheless, it is clear that investment to improve the basic education system will bring a high return on investment for the next generation and for sustainable development.

Secondary-school (college) education: Learning for what?

The secondary-school level will most probably be located in certain regional centres in the main valleys inside, or in the lowlands just outside, the mountains. The focus of secondary education cannot be limited to training students for admission to a university in order to increase the number of jobless graduates but must include an education and preparation for leading positions in all fields of mountain development. It therefore needs to have contact with the problems of the outside world and should include an introduction to modern communication technologies and a higher-level foundation for managing concrete development projects. The aim should be to provide an overall education to promote critical thinking, learning about interacting driving forces, thinking in integrated systems, planning in short- and long-term scales, cooperating with local people and policy makers, and working successfully in a team.

Tertiary (university) education: Responsibility for nature and society

Universities should rethink their objectives and their organization in order to offer the necessary courses in sustainability science and to stimulate research projects on preserving life-support systems, also for mountain areas. Universities are responsible for the education of teachers and for training a scientific community that is competent and capable of participating in the development process and communicating with the political authorities.

Scientists have a growing understanding of issues such as climate change, management of natural resources, population trends, consumption and waste, cultural diversity and conflicts, and environmental degradation. This knowledge should be used to shape long-term strategies for sustainable development. For this reason, capacity (and even institution) building for young scientists must be top priorities for developing countries.

The University of Central Asia, established in 2001 by an international treaty signed by the presidents of Tajikistan, Kyrgyzstan, and Kazakhstan and by His Highness the Aga Khan, intends to be the first university with teaching and research programmes dedicated exclusively to the problems and potentials of mountains and mountain peoples. It is a fascinating project and a real model for sustainable development in mountain areas. We hope that it will prove to be a great success for a very important and very critical mountain region (Hurni and Jansky 2001).

Capacity development

Capacity building has been very high on the political agenda of WSSD in Johannesburg and is an integral part of the process of education, the objective of which is to develop human potential to address the changes of human security and development. Institution building is much more demanding than capacity building, because it needs long-term engagement. However, we should not forget that the scientific communities in the richer countries have a responsibility to do much more for capacity and institution building in the developing world. What is the value of the top research results if they serve only the technological progress and the economic profit of the rich countries and the small élites in the poor countries, altogether perhaps only 20 per cent of the world's population? How can we speak about "global change" research programmes, if most of the world is not participating and an optimal transfer of important results to political authorities is not possible?

The scientific communities of the North have a responsibility to support the scientific communities in the South, not only with short-term capacity building but also with long-term engagement in institution building, so that the South can become self-sufficient. One example is provided by the mountain and cross-cutting issues-linked initiatives of the United Nations University (UNU 2002), which focus on strengthening national capacities in education and research. These initiatives assist in sustainable management of natural resources by providing support to scientists, promoting research networking, organizing training workshops, improving access to scientific information, and fostering partnership. The overall goal driving these activities is to (1) upgrade the knowledge, communication, and managerial skills necessary to address more effectively emerging issues in sustainable mountain development, and (2) promote information dissemination among local communities, policy makers, academics, researchers, and other institutions. This has the advantage not only of avoiding duplication but also of pooling and sharing scarce resources as a solution to the shortage of expertise in many countries of the South.

The main aim should be to persuade the political authorities in the South that economic development in the framework of sustainability is not possible without scientific knowledge and advice. True North–South research partnerships, based on more responsibility and new priorities of universities and science foundations in the North, while respecting local cultures and ensuring a new political and social esteem for science in the South, are indispensable for the future of our planet.

Modern information and communication technologies

Science-rich nations should be willing to share their knowledge with regions and countries that do not yet have the same infrastructure and scientific knowledge base. It is unrealistic to think that this will be possible in the near future for remote mountain areas; nevertheless, places of higher education should be connected to the outside world in order to use existing knowledge such as information libraries. One example is www.scidev.net, sponsored by major science publications and funded by development agencies, which will provide such resources free of charge to non-profit institutions in the 40 poorest countries and at substantial discount to those in other developing nations. The emphasis obviously will be on material relevant to developing countries. UNESCO is implementing activities for establishing Community Multimedia Centres, as a global strategy for addressing the digital divide at the local level for poor and underprivileged communities in the developing world (UNESCO 2002). This strategy complements efforts made at institutional, political, and regulatory levels. Other tools, such as geographical information systems (GISs) and global positioning systems (GPS) could be of interest for short- and longer-term field studies of, for instance, changes in land use and biodiversity. Any progress in the use of modern technologies should be promoted, even if there are many obstacles from the financial, technological – and, perhaps, even from the political – side, which must be overcome in the coming years.

Distance learning and open universities

New methods of learning and new institutions for the transfer of knowledge must be examined and should be initiated wherever possible. They could be of the greatest interest to mountain areas, but they depend on modern information technology. In the Highlands and Islands of Scotland, the emerging University of the Highlands and Islands links 13 academic partners and over 60 learning centres in dispersed communities, delivering further- and higher-education training and courses. In the highly dissected and isolated mountain areas of the developing world, the implementation of such high-cost initiatives may not be as easy; nevertheless, at least all the centres of higher education could be connected with these new communication technologies. A valuable model is the distance-learning course on watershed management offered by the University of British Columbia, and taken by people around the world. UNESCO's initiative on Community Multimedia Centres could also offer the necessary access to the computer, fax, and telephone in certain central locations in mountain regions. Distance learning and open universities could be interesting instruments for mountain development, even if an experimentation phase is still needed before more precise recommendations and instructions are defined (Schreier 2002, 2003).

Gender challenges

"Women in Sustainable Development" is a special chapter, No. 24, in Agenda 21 and there is also a Beijing Platform for Action. All these declarations show the importance of the topic, even if the progress in the ten years since Rio 1992 is extremely modest – or even non-existent in certain regions of the world.

The education of girls and the empowerment of women are key elements of survival and development in mountain areas. Only a few statistics, with disaggregated data about age and gender, are available as a basis for planning and policy evaluation. The full engagement of women in all aspects of life, in development projects, and decision-making processes, is essential for mountain communities.

Educational programmes and mountain cultures

Educational programmes can strengthen mountain cultures by demonstrating the scientific and practical value of ecological and medicinal knowledge possessed by elders and healers and by enlisting their services as teachers (Bernbaum 2000). They can also incorporate traditional ways of passing on knowledge and culture, many of which occur informally through work and apprenticeship rather than through classroom teaching. This is particularly important for nomadic and pastoral societies, where forcing children to stay in one place for schooling disrupts families and traditional ways of life. Some institutions offer education/capacity development programmes for local communities in conservation and sustainable development. For instance, the People, Land Management and Environmental Change (PLEC) programme of the University of the United Nations offers training courses for local communities, including those living in mountain regions.

Conclusions

Local communities and local authorities have key roles to play in making sustainable development happen. Without the integration of all people – men, women, and the local leaders – in a learning and decision process, any progress will be very difficult. Development concepts that analyse causes and effects of major issues are essential in education and school programmes from the local to the national level. If they are discussed in a local school, then they may also initiate a discussion and even a learning process among local people. This interaction between two generations, supported by the existing information media (radio and TV) and, in future, also by modern communication technologies, may help to create a learning society. Such a life-long education process is a key to find the balance between emigration from mountain communities to the valley and lowland centres – which has taken place for centuries in mountains around the world – and the continuation and stabilization of mountain communities and ecosystems. Education for sustainable development should be available to people of all ages and cultures.

Science

The role of science in the last two centuries

Exactly 200 years before the International Year of Mountains, Alexander von Humboldt initiated his fieldwork on Chimborazo (6,310 m), in Ecuador, which focused on understanding the ecology of the different altitudinal belts. His subsequent studies, concerning the vertically differentiated ecosystem belts in the tropical Andes; in the Himalayas; and along a transect linking Northern Scandinavia, the Alps, the Pyrenees, and the Pico de Teide on the Canary Islands, were published in the mid-nineteenth century (Bromme 1851). This was the beginning of modern mountain research involving worldwide comparisons and was a stimulus for numerous later studies. After about 1870, natural disasters in the European Alps helped to promote the natural and engineering sciences. From this time onwards, predominantly natural-science studies emerged in the mountains of the world; however, in general, they were considered quite marginal for the leading natural and social sciences and played a role only in connection with the interests of the economic, political, and cultural centres in the surrounding lowlands. This status changed after the 1970s, with rapidly growing interest in environmental problems (e.g. climate change), renewable and non-renewable natural resources (e.g. water, biodiversity, recreation), and mountain societies (e.g. cultural diversity, conflict solution, governance for sustainable development), all of which emerged as driving research themes in mountain regions (FAO 2000).

Natural, social, health, and engineering sciences: The need for inter- and transdisciplinarity

Disciplinary research in mountain regions has played a prominent role and will continue to do so. Among the many important disciplines are natural sciences (e.g. climatology), social studies (e.g. cultural–linguistic studies), health research (e.g. blood and lung problems), and engineering (e.g. flood prevention). However, the close relationship between natural processes and human activities in mountain areas has created a high demand for integrated natural–social science projects in order to furnish the knowledge necessary for addressing sustainable-development issues. Interdisciplinarity is the classical approach for scientific cooperation, but very often we praise interdisciplinarity and still promote disciplinarity. Transdisciplinarity is much more: it is a joint problem-solving among science, technology, and society (Thompson Klein et al. 2001). It takes up the real problems of the mountain world, including cooperating with practitioners and the integration of local people and communities in the process of knowledge production. The various stakeholders must participate from the beginning and be kept interested and active over the entire course of the project. Evaluation procedures and criteria must, therefore, be adapted to this approach, which should help to understand complexity, inform and integrate society, and prepare the relevant knowledge for political decisions and sustainable development. At the same time, it must be recognized that the effective implementation of trandisciplinarity remains challenging.

Research priorities for the future

Water resources

Mountains are the origin of much of the world's freshwater resources. However, given that water scarcity and even water crises are anticipated in certain regions of our planet during this century, our knowledge about the water cycle, especially in the mountains of the tropics and subtropics, is absolutely inadequate. The arid and semi-arid regions are most critical, as often more than 80–90 per cent of the available fresh water originates from mountains and uplands, and it is even more important to realize that, in these regions, 70–90 per cent is used for irrigation and food production. Water management for quantity and quality begins in the mountains; transdisciplinary projects are needed to integrate under-standing of natural variability and human interferences in highland–lowland systems, combining natural, social, health, and engineering sciences with the participation of local population and policy makers (Viviroli, Weingartner, and Messerli 2003).

Biodiversity, forests, protected areas, and land use

The different altitudinal belts of mountains represent a compression of different climatic zones along vertical gradients – one reason why they are unique hotspots of biodiversity. Protected areas can play important roles for the preservation of biodiversity, but we should not forget that carefully managed "cultural landscapes" also preserve high biological diversity (Körner and Spehn 2002). UNESCO's World Network of Bio-sphere Reserves, and other networks of protected areas in mountain re-gions around the world, could provide ideal test sites for research into, and monitoring of, global environmental change and its effects on bio-diversity and land use. In addition, high-energy mountain environments and associated gravity-driven processes mean that both people who live in mountains and those that travel through them require healthy forests for protection as well as production (Hamilton 1993). Consequently, we need to have a much greater knowledge of local and regional biodiversity in connection with climate variability, relief, soil and water conditions, changing land uses, social and demographic problems, tourism, and en-gineering projects in order to understand the vulnerability of mountain ecosystems and mountain biodiversity.

Land use and land degradation

Human impacts on mountain ecosystems have a long and well-documented history that allows us to compare periods of sustainable use and periods of deforestation and land degradation. In 400 BC, the Greek philosopher Plato wrote the following: "... it had much forest-land in its mountains ... what now remains, compared with what then existed, is like the skeleton of a sick man" (Löffler 1999). The over-exploitation of mountain resources in the Greek and Roman period can still be seen in some Mediterranean mountains, demonstrating the long-term nature of the costs associated with the destruction of natural capital such as soils and forests.

Economic standards, cultural diversity, natural resources, and sustainable development

The social and economic well-being of mountain communities, with their cultural diversity and identity, is a pre-condition for the sustainable use and management of mountain watersheds and resources, with ensuing benefits for the populations of surrounding lowlands (UNDP 1998). However, we must keep in mind that biogeophysical, social, cultural, economic, and political conditions are major components of this integrated system and that these often differ from one mountain system to another – or even between adjacent mountain valleys (Biosphere Reserve Integrated Monitoring [BRIM] 2002). It has been estimated that around 40 per cent of the mountain population in developing and transition countries are vulnerable to food insecurity. However, whereas most mountain people are rural (particularly in the Asia/Pacific region and sub-Saharan Africa), globally, 27 per cent of mountain people are urban, and settlements in and adjacent to mountain areas are expanding (Huddleston et al. 2003). Considerable work is needed to refine these figures and to understand the forces behind them – and their interactions.

Mountain cultures and scientific knowledge

Mountain cultures have an intimate knowledge of local environments culled from generations of experience. They know which plants are good for eating and which for healing; they know the habits of wildlife and where to find reliable sources of water and nourishment for their livestock. This traditional knowledge has many valuable uses, both for local communities and for the outside world. Much of it has neither been researched nor recorded by scientists and is in danger of being lost, as those entrusted with its keeping die out. Some of this knowledge – particularly that concerning sacred sites, substances, and practices – is privileged information and its secrecy needs to be respected (Bernbaum 2000; The Mountain Institute 1998).

Tourism, economic impacts, and cultural change

About 50 per cent (or even more) of the world population is living in urban areas, and we do not know yet how and when this process may come to certain limits. Coastal zones and mountain regions are the favoured potential recreation areas, with all the consequences on economies and cultures. For the mountains of the world, tourism is a special form of highland–lowland interaction. Few communities remain, or will remain, untouched by the presence of tourists and the industries that support them. At the same time, tourists themselves are confronted by aspects of economic impacts and cultural changes at every destination they visit. As one of the most visible mechanisms of globalization, tourism's magnitude and pervasiveness makes it a potent force in the process of cultural change. The challenge of sustainable development and the increasing recognition of the cultural dimension of biodiversity and landscape, the negative and positive impacts of tourism development, and patterns of tourist behaviour on culture, must remain important foci of academic attention. In this sense, research projects have to develop locally and regionally well-adapted strategies, showing how a low negative impact of visitors on nature and culture can be combined with a beneficially active socio-economic involvement of the local population (BRIM 2002).

Climate change, natural risks and disasters, human activities, and vulnerability of mountain systems

The highest ecosystem, above the timberline, is the only one that connects all the different climatic zones of the world in a pole–equator–pole transect. This ecosystem at the edges of the land–atmosphere interface – which hosts glaciers, snow, permafrost, and the uppermost limit of vegetation – is one of the most sensitive and globally comparable indicators of climate change. Even small changes in temperature and precipitation can produce natural hazards locally or, when reinforced by the kinetic energy of high-relief processes, disasters that affect adjacent lowlands. Human activities – including those associated with globalization – can also increase the vulnerability of mountain ecosystems. We therefore need coupled natural–human research strategies and models to detect and to predict the components that enhance or reduce the vulnerability of mountain ecosystems and mountain communities (Becker and Bugmann 2001) (fig. 11.1).

Highland–lowland interactions

"Highland–lowland interactions" is a concept that covers a broad range of processes and effects (Ives, Messerli, and Jansky 2002). Steep slopes imply geomorphologically high-energy environments, where atmospheric weathering processes combined with gravity can produce significant downslope mass transfer and, hence, inhibit the development of a diverse vegetation cover and mature soil profiles. Together with slope instability, this restricts biomass productivity and increases vulnerability to human intervention. Frequently, when downslope mass movements occur suddenly and involve large volumes of materials, they can be catastrophic to human life and property. They thus magnify the prevailing degree of inaccessibility or, alternatively, augment the maintenance costs of infrastructure extended into mountain regions in efforts to improve accessibility. This, in turn, causes high transport costs for goods imported by mountain communities and for products exported by them to potential lowland markets. Many of these problems have recently received an increasingly widespread and sympathetic reaction amongst aid agencies, with rapidly spreading realization that improved local access to resources in mountain areas will have widely beneficial effects in social, economic, and even political terms (Royal Swedish Academy of Sciences 2002).

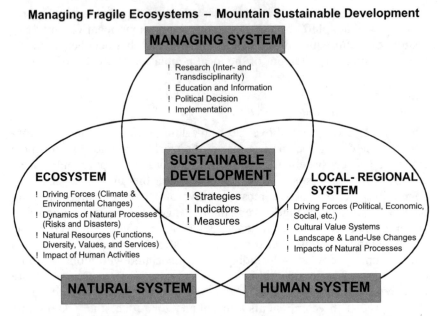

Figure 11.1 Understanding the complexity of mountain sustainable development

Science: A comment from Nepal

Until now, there has not been any significant effort for a true partnership be-tween North and South researchers, which is one reason why the complexity of highland–lowland linkages in the Himalayas is well known and understood by professors and researchers in the developed world, whereas concerned local sci-entists and professors remain at the mercy of information disseminated by the academics from the developed world. Unless there is sincerity and commitment of true collaboration, with a benefit for both parties, any notion of strengthening capabilities in the mountains of the developing world will never materialize (Sanjay Nepal).

Local case studies, regional knowledge centres, and global networks

Case studies at the local level will continue to play a fundamental role in future research programmes, because the complexity of interactions can be better analysed and understood in a well-defined area. Only then can we set priorities for theoretical modelling and practical application. The results of different case studies must be stored in a regional knowledge centre (e.g. ICIMOD for the Hindu Kush–Himalaya), in a regional node (e.g. those of the Mountain Forum), or in a special institution where the range of validity of the different case studies can be evaluated. However, this local and regional knowledge must be much better integrated at a global level, where we have the Global Change Research Programmes – the International Geosphere–Biosphere Programme (IGBP), International Human Dimensions Programme (IHDP), World Climate Research Programme (WCRP) and DIVERSITAS. These have advanced knowledge about global environmental change in a decisive way since Rio de Janeiro 1992. Mountain science has found its special place, even at this global level, with the following programmes:

- The Mountain Research Initiative, MRI (IGBP, IHDP, GTOS, UNESCO)
- The Global Mountain Biodiversity Assessment, GMBA (DIVERSITAS)
- The Global Observation Research Initiative in Alpine Environments, GLORIA (EU)
- Measuring and modelling the dynamic response of remote mountain-lake ecosystems to environmental change, a programme of Mountain Lake Research, MOLAR (IGBP–Past Global Changes (PAGES))
- World Glacier Monitoring Service, WGMS
- Mountain Biosphere Reserves, UNESCO Man and Biosphere programme
- International Hydrological Programme, UNESCO–IHP
- Global Terrestrial Observing System, GTOS: the Mountain Module (International Council of Scientific Unions (ICSU), FAO)
- Global Mountain Partnership Programme, UNU–GMPP.

Mountain-focused scientific journals such as *Mountain Research and Development*, and the Mountain Forum (www.mtnforum.org), also serve as a forum for mountain scientists to exchange information on research, views, and opinions.

Indicators, knowledge management, and sustainable development

Many highly aggregated economic indicators have been widely adopted, from the national to the global level. However, indicators concerning the environment and natural resources are missing in many countries and regions, especially in mountain areas. Data about the social, economic, and environmental conditions are, in many cases, lacking, and internationally published lists of indicators are often not adapted to mountain regions (OECD 2001). Moreover, we should keep in mind that these three groups of indicators must be complemented in mountain regions by specific components, including natural hazards, cultural identities, accessibility, and natural resources, among others. That aside, the Human

Development Index could be very helpful to compare different mountain regions and to indicate potential risks and crises (Kreutzmann 2001). Science not only should work in a retrospective sense, to tell us what went wrong in the past, but should also tell us what is doable, the longer-term strategic goals, and the possible scenarios that will allow us to reach these goals. In all these efforts, we should not forget that the pathway to sustainability cannot be charted in advance: it will have to be navigated through trial and error and conscious experimentation. Adaptive management and social learning are essential (US National Research Council 1999).

Scientific knowledge and policy making

The end of the twentieth and the beginning of the twenty-first century have seen growing concern about regional and global threats to humanity. These issues present an unprecedented challenge to scientists and policy makers alike, highlighting the importance of analysing and optimizing the interactions between the two groups (OECD 1998; see fig. 11.2). Two models for connecting scientific knowledge to policy-making can be described. The historic model involves two independent groups of players: the scientists analyse, interpret, evaluate, and report; the politicians decide, using their own competence, whether they do or do not wish to use the scientific advice. This procedure has never been satisfactory for solving highly complex problems. The newly proposed assessment model is based on three steps: the first one is scientific analysis,

Two models for connecting scientific knowledge to policy decision-making

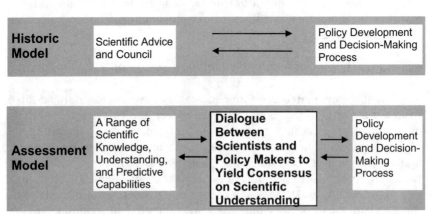

Figure 11.2 For a better dialogue between science and policy.
Source: OECD (1998)

which should not produce just a simple result but should show the whole range of scientific knowledge with its certainties, uncertainties, and predictive capabilities. The second step requires dialogue between scientists and policy makers in order to bring about consensus on scientific understanding (fig. 11.2). The politician must learn to understand complexity. However, the scientist has also to learn what can be done in the short term, which is the time-frame of interest to politicians – even if it is clear that important changes can never be reached in a few years. The process therefore requires, as a third step, a decision that goes in the right direction, so that a logical progression of decisions in the same line is possible in subsequent political periods. This phasing and mutual understanding of a political decision process is an important part of the dialogue between scientists and policy makers.

Conclusions

We need a new breed of scientists – scientists who understand, besides their speciality, how both natural and human systems operate and interact; scientists who can think locally, nationally, and globally. Perhaps this need will have to be met not only by a more efficient dialogue between scientists, politicians, and the local population, but also by restructuring some of our traditional scientific institutions in order to improve research and development in the mountains of the world.

REFERENCES

Becker, A., and H. Bugmann (eds). 2001. *Global change and mountain regions.* IGBP Report No. 49. Stockholm: The Mountain Research Initiative.

Bernbaum, E. 1997. *Pilgrimage and conservation in the Himalayas: A model for environmental action based on cultural and spiritual values.* Project report available from The Mountain Institute, Franklin, West Virginia.

Bernbaum, E. 1998. *Sacred mountains of the world.* Berkeley: University of California Press.

Bernbaum, E. 2000. "The cultural and spiritual significance of mountains as a basis for the development of interpretive and educational materials at national parks." *Parks* Vol. 10, no. 2.

BRIM. 2002. *Biosphere reserve integrated monitoring: Social monitoring.* Paris: UNESCO.

Bromme, T. 1851. *Atlas zu Alexander von Humboldt.* Stuttgart: Kosmos.

FAO. 2000. *International Year of Mountains.* Concept Paper. Rome: FAO.

Forsyth, A. 2001. Book review of Leat, D. and Nichols, A. 1999. *Theory into Practice, International Research in Geographical Environmental Education* Vol. 10, No. 2.

Hamilton, L. (ed). 1993. *Parks, peaks, and people*. Honolulu: East–West Center, Programme on Environment.

Huddleston, B., E. Ataman, P. de Salvo, M. Zanetti, M. Bloise, J. Bel, G. Francheschini, and L. Fè d'Ostiani. 2003. "Towards a GIS-based analysis of mountain environments and populations." Working paper No. 10. Rome: FAO.

Hurni, H., and L. Jansky. 2001. "Sustainable use and conservation of natural resources in the High Pamir Mountains of Central Asia." Work in progress. A review of research activities of the United Nations University. Vol. 16, No. 2.

Ives, J.D., B. Messerli, and L. Jansky. 2002. "Mountain Research in South-Central Asia: An overview of 25 years of UNU's Mountain Project." *Global Environmental Research* Vol. 6, No. 1.

Keating, M. 1993. *The Earth's Summit Agenda for Change. A plain language version of Agenda 21 and the other Rio Agreements*. Geneva: Centre of Our Common Future.

Kemf, E. (ed). 1993. *The law of the mother. Protecting indigenous peoples in protected areas*. San Francisco: Sierra Club.

Körner, C., and E. Spehn (eds). 2002. *Mountain biodiversity*. London: Parthenon.

Kreutzmann, H. 2001. "Development indicators for mountain regions." *Mountain Research and Development* Vol. 21, No. 2.

Löffler, H. 1999. "Die Wasservorräte – Ressourcen und Konflikte." In: H. Löffler and E. Streissler (eds) *Sozialpolitik und Oekologieprobleme der Zukunft*. Vienna: Oesterreische Akademie der Wissenschaft.

Messerli, B., and J.D. Ives (eds). 1997. *Mountains of the world: A global priority*. London: Parthenon.

Mountain Institute. 1998. *Sacred mountains and environmental conservation: A practitioner's workshop*. Harrisonburg, Virginia: Mountain Institute.

OECD. 1998. *Workshop on global-scale issues*, Stockholm, 4–6.3.1998, Paris: OECD.

OECD. 2001. *Environmental indicators. Towards sustainable development*. Paris: OECD.

Posey, D. (ed). 1999. *Cultural and spiritual values of biodiversity*. Nairobi: United Nations Environment Programme.

Royal Swedish Academy of Sciences. 2002. *The Abisko Agenda: Research for mountain area development*. Ambio Special Report No. 11. Stockholm: Royal Swedish Academy of Sciences.

Shengji, P. 1993. "Managing for biological diversity conservation in temple yards and holy hills. The traditional practices of the Xishuangbanna Dai Community, SW China." In: L. Hamilton (ed) *Ethics, religion and biodiversity*. Knapwell, Cambridge: White Horse Press.

Schreier, H. 2002. *Himalayan Andean Watershed Comparison*. 9 CD-ROMs. University of British Columbia, Vancouver, Canada: Inst. For Resources and Environment.

Schreier, H. 2003. *Integrated Watershed Management*. A graduate-level course offered over Internet. Vancouver: Institute for Resources and Environment, UBC. (http://www.ire.ubc.ca)

Stevens, S.F. 1993. *Claiming the high ground: Sherpas, subsistence, and environmental change in the highest Himalaya.* Berkeley: University of California Press.

Thompson Klein, J., W. Grossenbacher, R. Häberli, A. Bill, R.W. Scholz, and M. Welti (eds). 2001. *Transdisciplinarity: Joint problem solving among science, technology and society. An effective way of managing complexity.* Basle: Birkhäuser.

UNDP. 1998. *Trade and environment. Capacity building for sustainable development.* New York: UNCTAD.

UNESCO. 2001. *Proceedings of the Regional Thematic Expert Meeting on Potential Natural World Heritage Sites in the Alps.* Hallstadt, Austria, 18–22 June 2000. Vienna: text.um 4/01.

UNESCO. 2002. *UNESCO in the mountains of the world.* CD-ROM, UNESCO-UNEP-WCMC [also available at: http://valhalla.unep-wcmc.org/unesco/index.htm].

UNU. 2002. "UNU Public Forum – Mountains: Environment and human activities." CD-ROM. Tokyo: UNU Campus Computing Centre.

Viviroli, D., R. Weingartner, and B. Messerli. 2003. "Assessing the hydrological significance of the world's mountains." *Mountain Research and Development* Vol. 23, No. 1: 32–40.

US National Research Council. 1999. *A common journey. A transition towards sustainability.* Board on Sustainable Development. Policy Division. Washington D.C.: National Academic Press.

Appendix A

The Bishkek Mountain Platform

United Nations

 General Assembly

A/C.2/57/7

Distr.: General
7 November 2002

Original: English

Fifty-seventh session
Agenda item 86
Sustainable development and international economic cooperation

Letter dated 6 November 2002 from the Permanent Representative of Kyrgyzstan to the United Nations addressed to the Secretary-General

I have the honour to forward herewith the text of the Bishkek Mountain Platform which was formulated during the Bishkek Global Mountain Summit (28 October-1 November 2002) (see annex).

I would be grateful if you would circulate the text of the Bishkek Global Mountain Platform as a document of the General Assembly, under agenda item 86.

(*Signed*) Kamil **Baialinov**
Permanent Representative
of the Kyrgyz Republic to the United Nations

02-67947 (E)   111102
0267947

A/C.2/57/7

Annex to the letter dated 6 November 2002 from the Permanent Representative of Kyrgyzstan to the United Nations addressed to the Secretary-General

[Original: English and Russian]

Bishkek Global Mountain Summit
28 October–1 November 2002

Bishkek Mountain Platform

1.   Objectives

The Bishkek Mountain Platform is an outcome of the Bishkek Global Mountain Summit, the culminating global event of the International Year of Mountains 2002. The objective of the Platform is to continue with existing initiatives and to develop substantive efforts beyond the Year by mobilizing resources, giving orientation and guidance, and promoting synergies. In particular, it will provide a framework for stakeholders and others to contribute to sustainable development in the world's mountain regions. It will enable them to act together at all levels from local to global to improve the livelihoods of mountain people, to protect mountain ecosystems and to use mountain resources more wisely. The Platform should, furthermore, serve as a contribution to debate in the General Assembly of the United Nations and to the achievement of the Millennium Goals[1].

2.   Background

The Bishkek Mountain Platform builds on the rich experience embodied in documents on sustainable mountain development, beginning with Chapter 13, "Managing Fragile Ecosystems: Sustainable Mountain Development", of Agenda 21 of the United Nations Conference on Environment and Development in Rio de Janeiro in 1992. The ensuing process culminated in the International Year of Mountains, which was initiated by the Government of the Kyrgyz

---

- [1] To eradicate extreme poverty and hunger: the stated overall goal is to reduce the proportion of people living on less than $1 a day to half the 1990 level by 2015 — from 29 percent of all people in low and middle income economies to 14.5 percent. If achieved, this would reduce the number of people living in extreme poverty to 890 million (or to 750 million if growth stays on track)
- To achieve universal primary education
- Promote gender equality and empower women
- Reduce child mortality
- Improve maternal health
- Combat HIV/AIDS, malaria, and other diseases
- Ensure environmental sustainability
- Build a global partnership for development

A/C.2/57/7

Republic. The objectives of the Year are to "promote the conservation and sustainable development of mountain regions, thereby ensuring the wellbeing of mountain and lowland communities." In preparation for, and during, the Year, many meetings on different aspects of sustainable mountain development have been held (cf. list in Annex 1),* and their resolutions and declarations have also contributed to the Platform. A series of thematic papers, prepared for the Bishkek Global Mountain Summit by international specialists and developed further through electronic consultations, have also contributed to the Platform. Furthermore, it takes into consideration the recommendations of paragraph 40 of the Plan of Implementation of the World Summit on Sustainable Development in Johannesburg in August 2002.

3.   Challenges

Mountain areas cover 26 percent of the Earth's land surface and host 12 percent of its people. Mountains provide vital resources for both mountain and lowland people, including fresh water for at least half of humanity, critical reserves of biodiversity, food, forests and minerals. They are culturally rich and provide places for the physical and spiritual recreation of the inhabitants of our increasingly urbanised planet.

The people of mountain areas face major challenges. About half of the world's approximately 700 million mountain inhabitants are vulnerable to food shortages and chronic malnutrition. Mountain people, particularly disadvantaged groups such as women and children, suffer more than others from the unequal distribution of assets and from conflicts.

Policy decisions influencing the use of mountain resources are generally made in centres of power far from mountain communities, which are often politically marginalized and receive inadequate compensation for mountain resources, services and products. Mountain ecosystems are exceedingly diverse but fragile because of their steep slopes, altitude and extreme landscapes. Many of these ecosystems are being degraded because farmers are forced to apply unsustainable agricultural practices and by inappropriate development.

Climate change, natural hazards and other forces also threaten the complex webs of life that mountains support. The consequences of poverty and environmental degradation reach far beyond mountain communities, through war, terrorism, refugee migration, loss of human potential, drought, famine, and escalating numbers of landslides, mudslides, catastrophic floods and other natural disasters in highlands and lowlands. Moreover, the rapid melting of mountain glaciers and degradation of watersheds is reducing the availability of life-sustaining water and increasing the potential of conflict over dwindling supplies.

---

* Annex not included.

A/C.2/57/7

4. Declaration

We, the participants in the Bishkek Global Mountain Summit, the culminating global event of the International Year of Mountains, pledge our long-term commitment and determination to achieving the goals of sustainable development in mountain areas. We are committed to protecting the Earth's mountain ecosystems, eliminating poverty and food insecurity in mountain areas, promoting peace and economic equity, and providing support for current and future generations of mountain people – women and men, girls and boys – to create the conditions in which they can shape their own goals and aspirations.

5. Guiding principles

We support participatory, multi-stakeholder, multi-disciplinary, eco-regional, decentralized and long-term approaches that respect the principles of subsidiarity, human diversity, human rights, gender equity and the environment. We value and build upon both indigenous and scientific information and knowledge.

6. Framework for action

We call on the United Nations and its organizations, countries, international and non-governmental organizations, businesses, grassroots organizations, scientists and individuals to jointly invest their resources in mountain areas. We also call on financial institutions, including the GEF, to continue and increase their support. It will take all of us, working in partnership, to achieve our goals. We see this framework as guidance for coming decades, recognizing that the details will be developed by partners.

6.1. Actions at the international level

United Nations Resolution:
We suggest that the International Year of Mountains Focus Group of the United Nations develop a United Nations resolution on sustainable development in mountain regions. The resolution might provide guidance for the United Nations and its agencies to develop policies and programmes in accordance with the objectives and principles of the Platform, and invite further cooperation and enhancement of actions in mountain regions worldwide. Furthermore, we encourage the Focus Group to highlight the vital inter-relationships between mountains and freshwater resources, particularly in the context of the International Year of Freshwater 2003, and to consider the establishment of a World Mountain Day.

International Partnership:
We support the International Partnership for Sustainable Development in Mountain Regions, a 'type 2' outcome of the World Summit on Sustainable Development in Johannesburg in August 2002. We welcome the offer of FAO to host the secretariat of the Partnership and bring the Inter-Agency Working Group on Mountains to its service. We call on UNEP to ensure

A/C.2/57/7

environmentally sound management in mountain regions, in particular in developing countries, by strengthening environmental networking and assessments, facilitating regional agreements and encouraging public-private sector cooperation. We count on the continuing and increasing involvement of UNDP, UNESCO, UNU, other United Nations agencies, multilateral development banks, other international organizations and countries.

The structure and working modalities will be further elaborated to ensure an effective Partnership. We invite interested organizations and countries to join the Partnership and ensure its financial sustainability.

We welcome the proposal to create, within the context of the Partnership, an international Network of Developing Mountain States and Regions, and support the establishment of a working group for its further elaboration.

Capacity development:
We believe that capacity development at all levels is essential to improve the competence of mountain stakeholders and to enhance understanding of mountain processes, problems, needs, opportunities and assets. This should involve all sectors of education, NGOs, governments, decision-makers and international agencies.

Science and technology:
We invite the scientific community and its funding agencies, at international and national levels, to promote international partnerships and programmes of research, monitoring and early warning in support of sustainable development in mountain regions. We particularly emphasize that initiatives should focus on biophysical as well as political, social, economic and cultural aspects, and that they apply disciplinary, interdisciplinary and transdisciplinary approaches, thereby contributing to integrated understanding of problems and opportunities for sustainable mountain development.

6.2. Actions at the regional (supra-national) level

Regional focus:
We are convinced that transboundary mountain regions have specific environmental, social, political, cultural and economic characteristics and potential for development and therefore require specific approaches and resources.

Regional cooperation:
We urge that development and conservation in transboundary mountain regions and between upstream and downstream stakeholders be coordinated between all partners affected or involved.

A/C.2/57/7

Regional agreements:
We support formal instruments such as charters, conventions and integrated policies to foster international cooperation between states sharing mountain areas.

6.3. Actions at the national level

Governance:
We call upon national governments to apply the principle of subsidiarity by delegating political decisions to the lowest possible level of decision-making, from national to sub-national to community, corporate and private responsibilities.

Policy advocacy:
We invite national governments to develop legislation, policies and procedures in favour of their mountain areas, particularly those that are marginalized in terms of economic and social development, and to set their national priorities accordingly. We also invite political parties and governments to become involved with international initiatives, provided that these are accepted at the local level.

Mountain-specific data:
We recognize that the lack of spatially disaggregated socio-economic and environmental data hampers the recognition and specific analysis of mountain livelihood issues. We encourage governments to produce, publish and use mountain-specific data to improve policies for sustainable mountain development, especially in relation to dominant lowland economies.

Investment and compensation mechanisms:
We are convinced that economic disparities between mountains and their surrounding areas can be reduced through investment and other means. We encourage governments to introduce compensation mechanisms for goods and services provided by mountain communities, enterprises or natural and cultural landscapes through negotiations between affected people and beneficiaries.

Providing access:
We recognize that the physical nature of mountain regions hinders access in many ways. In particular, we call upon governments to use information and communications technologies to bring benefits to mountain people.

6.4. Actions at the local level

Local stewardship:
We support local governance and ownership of resources, individual freedom, cultural self-determination, and traditional belief systems, which lie at the core of sustainable development in mountain areas, especially where the economic influence of external forces is high.

A/C.2/57/7

Local development:
We urge all stakeholders to ensure that local livelihoods are improved, economic enterpreneurship is fostered and that environmental protection and sustainable use of natural resources is guaranteed. External partners should seek to support local initiatives when requested to do so.

Appendix B

UN Resolution UN GA A/Res/57/ 245 from 20 December 2002

United Nations   A/RES/57/245

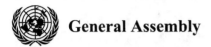 General Assembly

Distr.: General
30 January 2003

Fifty-seventh session
Agenda item 86

Resolution adopted by the General Assembly

[*on the report of the Second Committee (A/57/531/Add.5)*]

57/245.   International Year of Mountains, 2002

The General Assembly,

Recalling its resolution 53/24 of 10 November 1998, in which it proclaimed 2002 as the International Year of Mountains,

Recalling also its resolution 55/189 of 20 December 2000,

Recognizing chapter 13 of Agenda 21[1] and all relevant paragraphs of the Plan of Implementation of the World Summit on Sustainable Development ("Johannesburg Plan of Implementation"),[2] in particular paragraph 42 thereof, as the overall policy frameworks for sustainable mountain development,

Noting the voluntary International Partnership for Sustainable Development in Mountain Regions, launched during the World Summit on Sustainable Development with the committed support of twenty-nine countries, sixteen intergovernmental organizations and sixteen organizations from the major groups, as an important approach to addressing the various interrelated dimensions of sustainable mountain development,

Taking note of the Bishkek Mountain Platform, the outcome document of the Bishkek Global Mountain Summit, held at Bishkek from 28 October to 1 November 2002, which was the concluding event of the International Year of Mountains,

1.   *Takes note* of the interim report transmitted by the Secretary-General on the International Year of Mountains, 2002;[3]

2.   *Welcomes* the success achieved during the International Year of Mountains, during which numerous activities and initiatives were undertaken at all levels, including major international meetings held in Bhutan, Canada, Ecuador, Germany, India, Italy, Kyrgyzstan, Nepal, Peru and Switzerland, which catalysed a

---

[1] *Report of the United Nations Conference on Environment and Development, Rio de Janeiro, 3–14 June 1992* (United Nations publication, Sales No. E.93.I.8 and corrigenda), vol. I: *Resolutions adopted by the Conference*, resolution 1, annex II.

[2] *Report of the World Summit on Sustainable Development, Johannesburg, South Africa, 26 August–4 September 2002* (United Nations publication, Sales No. E.03.II.A.1 and corrigendum), chap. I, resolution 2, annex.

[3] A/57/188.

02 55558

A/RES/57/245

strengthened interest for sustainable development and poverty eradication in mountain regions;

3.   *Recommends* that the experience gained during the International Year of Mountains be valued in the context of an appropriate follow-up;

4.   *Notes with appreciation* the effective role played by Governments, as well as major groups, academic institutions and international organizations and agencies, in the activities related to the International Year of Mountains, including the establishment of seventy-four national committees;

5.   *Also notes with appreciation* the work undertaken by the Food and Agriculture Organization of the United Nations as the lead agency for the International Year of Mountains, as well as the valuable contributions made by the United Nations Environment Programme, the United Nations University, the United Nations Educational, Scientific and Cultural Organization, the United Nations Development Programme and the United Nations Children's Fund;

6.   *Encourages* Governments, the United Nations system, the international financial institutions, the Global Environment Facility, within its mandate, and all relevant stakeholders from civil society organizations and the private sector to provide support, including through voluntary financial contributions, to the local, national and international programmes and projects resulting from the International Year of Mountains;

7.   *Invites* the international community and other relevant partners to consider joining the voluntary International Partnership for Sustainable Development in Mountain Regions;

8.   *Notes* that all stakeholders in the voluntary International Partnership for Sustainable Development in Mountain Regions have initiated a consultative process, with a view to determining the best options for further assisting stakeholders in the implementation of the Partnership, including through consideration of the offer made by the Food and Agriculture Organization of the United Nations to host a secretariat financed through voluntary contributions;

9.   *Encourages* all relevant entities of the United Nations system, within their respective mandates, to continue their constructive collaboration in the context of the follow-up to the International Year of Mountains, taking into account the inter-agency group on mountains, and the need for the further involvement of the United Nations system, in particular the Food and Agriculture Organization of the United Nations, the United Nations Environment Programme, the United Nations University, the United Nations Development Programme, the United Nations Educational, Scientific and Cultural Organization and the United Nations Children's Fund, international financial institutions and other relevant international organizations, consistent with the mandates specified in the Bishkek Mountain Platform;

10.   *Decides* to designate 11 December as International Mountain Day, as from 11 December 2003, and encourages the international community to organize on this day events at all levels to highlight the importance of sustainable mountain development;

11.   *Requests* the Secretary-General to submit to the General Assembly at its fifty-eighth session a report on the achievements of the International Year of

A/RES/57/245

Mountains, under a sub-item entitled "Sustainable mountain development" of the item entitled "Environment and sustainable development".

78th plenary meeting
20 December 2002

Appendix C

The International Partnership for Sustainable Development in Mountain Regions

INTERNATIONAL PARTNERSHIP FOR SUSTAINABLE DEVELOPMENT IN MOUNTAIN REGIONS

GUIDING PRINCIPLES

As approved by the Members of the International Partnership for Sustainable Development in Mountain Regions, commonly known as 'the Mountain Partnership', in preparation for their first global meeting in Merano, Italy, 5-6 October 2003.

BACKGROUND AND CONTEXT

For more than a decade the international community and the United Nations have increasingly focussed attention on improving livelihoods and environmental conditions in mountain regions throughout the world through the implementation of Chapter 13 of Agenda 21: "Managing Fragile Ecosystems - Sustainable Mountain Development".

The United Nations General Assembly (UNGA) declared 2002 as the International Year of Mountains to increase awareness of the urgent need to protect the world's mountain ecosystems and to improve the well-being of mountain people.

The importance of mountains and mountain ecosystems was strengthened by the launch of the International Partnership for Sustainable Development in Mountain Regions (herein after referred to as "the Mountain Partnership") at the World Summit on Sustainable Development (WSSD) Johannesburg, September 2002. The Partnership is a 'Type II' outcome of the Summit, and it aims at fostering the implementation of paragraph 42 of the WSSD Plan of Implementation, requiring actions at all levels to protect mountain environments and support mountain livelihoods through the integration of environmental, economic and social components of sustainable mountain development.

The Mountain Partnership is formally defined by its basic concept paper (herein after referred to as the "Bali Document"), which was discussed and finalized by interested parties during the preparatory process for the WSSD.

At the end of 2002, the significance of mountains was further recognized in UNGA Resolution for the International Year of Mountains (A/RES/57/245). The UNGA Resolution also took note of the Bishkek Global Mountain Platform, a framework for future action for sustainable mountain development, which was issued at the Bishkek Global Mountain Summit (November 2002), the culminating event of the International Year of Mountains.

The initiators of the Mountain Partnership (the Government of Switzerland, FAO, UNEP), as well as the Government of Italy and the Mountain Forum, have since begun a process to pursue the further development of the Partnership. This iterative process has so far included a series of key steps in 2003, including: two meetings in Switzerland, an e-consultation moderated by the Mountain Forum, open to all members of the Mountain Partnership, a side event on the occasion of CSD-11 in New York, as well as the preparatory process for the first global meeting of the members of the Mountain Partnership in Merano, Italy.

OBJECTIVES AND STRATEGY

In accordance with the general goals as defined in the Bali Document, the Mountain Partnership is a voluntary alliance of national governments, intergovernmental organizations, local and regional authorities, non-governmental organizations, the private sector, the academic community and other major group representatives who are working together to "improve livelihoods, conservation and stewardship throughout the world's mountain landscapes". The goals and priority areas are those listed in paragraph 42 of the WSSD Plan of Implementation.

The Mountain Partnership will encourage and promote concrete initiatives and alliances, e.g. in the form of participatory programmes and projects, at global, national and local level which protect mountain ecosystems, safeguard social and cultural traditions and address chronic poverty in mountain regions. An important principle in this context will be to make full use of already existing relevant networks and institutions and to learn and benefit from their experience with a view to promoting synergies, effectiveness and efficiency.

The Mountain Partnership will also contribute to the definition and implementation of policies based on an adequate assessment of natural and cultural resources of mountain populations and support the institutional capacities related to mountain ecosytems. It could also consider other themes like, e.g. mountains as a source of recreation and inspiration, sustainable energy production and use, reducing vulnerability to natural disasters, traditional knowledge and scientific research. Given the transboundary character of many mountain areas, it will be essential to link local, national and global efforts for long-term sustainable development.

STRUCTURE AND FUNCTIONS

As stated in the Bali Document, the Mountain Partnership is conceived as an umbrella alliance under which all partners can enter into specific initiatives according to their interest, competence and willingness. These initiatives constitute the building blocks of the Mountain Partnership and provide its concrete substance.

The Mountain Partnership will serve as a mechanism for networking, communication and information sharing and function as a clearing house for members. It will also complement, support and strengthen on-going initiatives in sustainable mountain development.

The Mountain Partnership will function as a broker for joint initiatives, facilitating contacts between countries and institutions in view of joint activities and creating conditions for cooperation and resource mobilization at the national, regional and global level.

The Mountain Partnership will forge linkages with existing multilateral instruments such as the Convention on Biological Diversity (CBD), the UN Convention to Combat Desertification (UNCCD), the UN Framework Convention on Climate Change (UNFCCC), the International Strategy for Disaster Reduction (ISDR) and other relevant instruments.

The Bali Document stipulates that the Mountain Partnership ought to have a common, easily accessible networking point (herein after referred to as "the Secretariat"). The Secretariat of the Mountain Partnership will help to link existing activities in mountain development and foster synergies and complementarities to promote closer collaboration, avoid duplication and achieve greater coherence and better results in terms of sustainable development. It will however not coordinate the actions of the members nor assign specific tasks and responsibilities to members.

The Secretariat will also be responsible for reporting to the Commission on Sustainable Development (CSD).

The Secretariat of the Mountain Partnership will be hosted by the Food and Agriculture Organization of the United Nations (FAO) and financed through voluntary contributions. It will be multi-stakeholder in composition. Its main tasks will be to provide communication services, knowledge management and brokering functions, acting as networking point and liaison office for partners.

During an interim period ending in 2004, a temporary arrangement to provide secretariat services has been established at FAO, with inputs provided by FAO, UNEP and the Mountain Forum. During this phase, it will assess the needs of the members. At the end of this interim period the work, structure and organization of the Secretariat will be reviewed and refined in a longer-term perspective.

GOVERNANCE

The Mountain Partnership is an evolving and flexible network of parties committed to working together. Its innovative nature as an umbrella partnership means that it does not require a complex governance mechanism. However, there is a need to guide the Mountain Partnership by a simple governance structure to accommodate stakeholders' interests and concerns and to ensure the legitimacy and transparency of decision-making.

The future governance of the Mountain Partnership will be based on principles, such as: democratic participation of all members; transparency and accountability; responsiveness; effectiveness; and efficiency.

The governance structure of the Mountain Partnership will be further defined through a consultative process open to all members.

MEMBERSHIP CRITERIA AND COMMITMENTS

Membership of the Mountain Partnership is open to all governments, including local and regional authorities, intergovernmental and other organizations whose objectives and activities are consistent with the vision and mission of the Mountain Partnership, and who fulfill the criteria for membership.

The criteria for membership of the Mountain Partnership include:

- Endorsement of the general principles and goals of the Mountain Partnership;
- Involvement in sustainable mountain development;
- Institutional establishment, with some level of stability in terms of funding and organizational long-term viability;
- Capacity to fulfil the membership roles and responsibilities, as defined by the Mountain Partnership.

Members of the Mountain Partnership will be expected to fulfill the following core roles and responsibilities:

- Initiating and participating in collaborative activities with other partners;
- Contributing to the Mountain Partnership activities by sharing information and knowledge through various channels, such as: providing links to the Mountain Partnership Web site; contributing success stories, case studies and/or lessons learned to the Mountain Partnership knowledge base; participating in virtual discussions and electronic conferences;
- Attending and contributing to Mountain Partnership meetings and events, wherever possible.

Rights and obligations of members of the Partnership will be further reviewed during the inception period and if necessary refined.

Acronyms

ACAP	Annapurna Conservation Project
AI	Appreciative Inquiry
ALPEX	Alpine Meteorology Experiment
APP	Agricultural Perspective Plan
APPA	Appreciative Participatory Planning and Action
BGMS	Bishkek Global Mountain Summit
BMP	Bishkek Mountain Platform
BNP	Bieszczady National Park
BPP	Business Promotion Project
BRIM	Biosphere Reserve Integrated Monitoring
CDM	Clean Development Mechanism
CGIAR	Consultative Group on International Agricultural Research
COP	Conference of the Parties
CRED	Centre for Research on the Epidemiology of Disaster
CRP	Conservation Reserve Program
CVC	Cauca Valley Corporation
DFID	Department for International Development
DNEF	Direction Nationale des Eaux et Forêts [National Directorate of Water and Forests]
EANET	East Asian Network on Acid Depositions
ECOSOC	UN Economic and Social Council
EIA	environmental impact assessment
FAO	Food and Agriculture Organization
FONAFIFO	National Forest Office and National Fund for Forest Financing (Costa Rica)

GDP	gross domestic product
GEF	Global Environment Facility
GIS	geographical information system
GLOF	glacier lake outburst flooding
GLORIA	Global Observation Research Initiative in Alpine Environments
GMBA	Global Mountain Biodiversity Assessment
GMPP	Global Mountain Partnership Programme
GPS	global positioning systems
GRC	Green Road Concept
GTOS	Global Terrestrial Observing System
HDI	Human Development Index
HIV/AIDS	human immunodeficiency virus/acquired immunodeficiency syndrome
IAGM	Inter-Agency Group on Mountains
ICDP	Integrated Conservation and Development Projects
ICIMOD	International Centre for Integrated Mountain Development
ICSU	International Council of Scientific Unions
ICT	information and communication technologies
IGBP	International Geosphere–Biosphere Programme
IHDP	International Human Dimensions Programme
IHP	International Hydrological Programme
ILO	International Labour Organization
ILRI	International Livestock Research Institute
INDOEX	Indian Ocean Experiment
INGOs	international non-governmental organizations
IPCC	Intergovernmental Panel on Climate Change
IRS	Indian Remote Sensing Satellite
IUCN	International Union for the Conservation of Nature
IYN	International Year of Mountains
KCC	Kangchendzonga Conservation Committee
KEEP	Kathmandu Environmental Education Project
KKH	Karakorum Highway (Pakistan–China Friendship Highway)
LANDSAT	Land Observation Satellite
LHWP	Lesotho Highlands Water Project
LISS	Linear Imaging and Self-Scanning Sensor
MCST	Mountain Conservation Society of Tanzania
MEAs	multilateral environment agreements
MFR	Makiling Forest Reserve
MOLAR	Mountain Lake Research
MRFF	Macquarie River Food and Fibre
MRI	Mountain Research Initiative
NEAP	National Ecotourism Accreditation Programme
NGOs	non-governmental organizations
NTFP	non-timber forest product
NYC	New York City
OFDA	Office of the US Foreign Disaster Assistance

PAGES	Past Global Changes
PES	payments for environmental services
PLA	Participatory Learning and Action
PLEC	People, Land Management and Environmental Change
RECOFTC	Regional Community Forestry Training Centre
RSA	Republic of South Africa
SF	State Forests of New South Wales
SLC	Snow Leopard Conservancy
SMD	sustainable mountain development
SMT	sustainable mountain tourism
SPOT	Système Probatoire d'Observation de la Terre
TAAN	Trekking Agents Association of Nepal
TAAS	Trekking Agents Association of Sikkim
TAWG	Tanga Aids Working Group
TIES	The International Ecotourism Society
TM	Thematic Mapper
TMI	The Mountain Institute
UNCED	United Nations Conference on Environment and Development (Rio 1992)
UNEP	United Nations Environment Programme
UNESCO	United Nations Educational, Scientific, and Cultural Organization
UPLB	University of the Philippines, Los Baños
USAID	United States Agency for International Development
USDA	US Department of Agriculture
USEPA	US Environmental Protection Agency
VEC	Village Electrification Committee
WAC	Watershed Agricultural Council
WBGU	Wissenschlaftlicher Beirat der Bundesregierung Globale Umwelt-verändeerung
WCED	World Commission on Environment and Development
WCRP	World Climate Research Programme
WGMS	World Glacier Monitoring Service
WHO	World Health Organization
WSSD	World Summit on Sustainable Development
WWF	Worldwide Fund for Nature
XS	multispectral

List of contributors

Edwin Bernbaum is the Director of the Sacred Mountains Program at The Mountain Institute, 1846 Capistrano Ave. Berkeley, CA 94707, USA; e-mail: ebernbaum@mountain.org

Wolfgang E. Burhenne is a Member of the Steering Committee (Liaison to the UN system) of the IUCN Commission on Environmental Law, and is also Executive Governor of the International Council of Environmental Law, Godesberger Allee 108-112, 53175 Bonn, Germany

Hans Van Ginkel is Rector of the United Nations University, 5-53–70 Jingumae, Shibuya-ku, Tokyo 150–8925, Japan; e-mail: rector@hq.unu.edu

Thomas Hofer is Forestry Officer, Sustainable Mountain Development, of the Forestry Department, FAO, Viale delle Terme di Caracalla, 00100 Rome, Italy.

Hans Hurni is Co-Director of the Centre for Development and Environment (CDE) at the University of Berne, Steigerhubelstrasse 3, 3008 Bern, Switzerland; e-mail: hans.hurni@cde.unibe.ch

Andrei Iatsenia is former Executive Secretary of the technical secretariat responsible for preparation of the Bishkek Global Mountain Summit; e-mail: andrei.iatsenia@weforum.org

Mylvakanam Iyngararasan is Senior Programme Officer at the UNEP Regional Resource Center for Asia and the Pacific, Asian Institute of Technology, P.O. Box 4, Klong-Luang 12120, Thailand; e-mail:

Mylvakanam.Iyngararasan@rrcap.unep.org

Libor Jansky is Senior Academic Programme Officer, Environment and Sustainable Development Programme, United Nations University, 5-53–70 Jingumae, Shibuya-ku, Tokyo 150 8925, Japan, jansky@hq.unu.edu

Walter Kahlenborn is Managing Director of Adelphi Research, gGmbH, Caspar-Theyss-Str. 14a, 14193 Berlin.

Andreas Kläy is a Co-Director of the Centre for Development and Environment (CDE), University of Berne, Steigerhubelstrasse 3, 3008 Bern, Switzerland.

Maritta R.v. Bieberstein Koch-Weser is President of Earth 3000, Palais am Festungsgraben, 10117 Berlin-Mitte, Germany.

Thomas Kohler is a Co-Director of the Centre for Development and Environment (CDE), University of Berne, Steigerhubelstrasse 3, 3008 Bern, Switzerland.

Wendy Brewer Lama is an Eco-tourism Consultant, 699 Spindrift Way, Half Moon Bay, CA 94019 USA; e-mail: wendylama@coastside.net

Douglas McGuire is Senior Forest Conservation Officer and Head of the Mountain Group, Forestry Department, FAO, Viale delle Terme di Caracalla, 00100 Rome, Italy.

Ali Mekouar is Chief of the Development Law Service, Legal Office, FAO, Viale delle Terme di Caracalla, 00100 Rome, Italy; e-mail: ali.mekouar@fao.org

Bruno Messerli is an[?OK] Emeritus Professor of the Geographical Institute, University of Berne, Hallerstrasse 12, 3012 Bern, Switzerland; e-mail: bmesserli@bluewin.ch

P.K. Mool is a Remote Sensing Analyst in Water, Hazard and Environmental Management at the International Center for Integrated Mountain Development (ICIMOD), 4/80 Jawalakhel, G.P.O. Box 3226, Kathmandu, Nepal; e-mail: mool@icimod.org.np

Safdar Parvez is Programs Officer, Asian Development Bank, Pakistan Resident Mission, Overseas Pakistanis Foundation (OPF) Building, Sharah-e-Jamhuriayat, G-5/2, Islamabad, Pakistan.

D. Jane Pratt is a Partner, EcoLogica, LLC., 36913 Paxson Rd., Purcellville, VA 20132, USA; e-mail: dpratt@mountain.org

Martin F. Price is Director, Centre for Mountain Studies, Perth College, UHI Millennium Institute, Crieff Road, Perth PH1 2NX, UK.

Stephen F. Rasmussen is a member of the Pakistan Microfinance Network, Block 14, Civic Centre, G-6, Islamabad, Pakistan; e-mail: srasmussen@worldbank.org

Maho Sato is an Associate Professional Officer in the Forestry Department, FAO, Viale delle Terme di Caracalla, 00100 Rome, Italy.

Nikhat Sattar is Head, Emerging and Emergency Programmes, IUCN Asia Programme (Sub-Office: 1, Bath Island Road, Karachi,

Pakistan; Main Office: 63, Soi 39, Sukhumvit, Bangkok, Thailand); e-mail: nikhat.sattar@iucnp.org

Surendra Shrestha is Regional Director and Representative, UNEP for Asia and the Pacific (UNEP ROAP), United Nations Building, Rajdamnem Avenue, Bangkok 10200, Thailand; e-mail: surendra.shrestha@rrcap.unep.org

Frederick Starr is Chairman of the Central Asia Caucasus Institute at the Paul H. Nitze School of Advanced International Studies (SAIS), Johns Hopkins University, Washington DC 20036, USA; e-mail: sfstarr@jhu.edu

Li Tianchi is a Senior Associate Scientist in Water, Hazard and Environmental Management at the International Center for Integrated Mountain Development (ICIMOD), 4/80 Jawalakhel, G.P.O. Box 3226, Kathmandu, Nepal.

Klaus Toepfer is Executive Director of the United Nations Environment Programme, United Nations Avenue, Gigiri, P.O. Box 30552, Nairobi, Kenya; e-mail: Klaus. Toepfer@unep.org

Annie Villeneuve is a Legal Consultant at the Legal Office, FAO, Viale delle Terme di Caracalla, 00100 Rome, Italy.

Teiji Watanabe is an Associate Professor in the Graduate School of Environmental Earth Science, Hokkaido University, N-10, W-5, Sapporo, Hokkaido 060-0810, Japan; e-mail: twata@ees.hokudai.ac.jp

Urs Wiesmann is a Co-Director of the Centre for Development and Environment (CDE), University of Berne, Steigerhubelstrasse 3, 3008 Bern, Switzerland.

Masatoshi Yoshino is an Emeritus Professor of the University of Tsukuba, and Senior Academic Advisor to the Environment and Sustainable Development Programme, United Nations University, 5-53–70 Jingumae, Shibuya-ku, Tokyo 150–8925, Japan; e-mail: yoshino@hq.unu.edu

Index

Catalogue Request

Name: _____

Address: _____

_____

Tel: _____

Fax: _____

E-mail: _____

To receive a catalogue of UNU Press publications kindly photocopy this form and send or fax it back to us with your details. You can also e-mail us this information. Please put "Mailing List" in the subject line.

 **United Nations University Press**

53-70, Jingumae 5-chome
Shibuya-ku, Tokyo 150-8925, Japan
Tel: +81-3-3499-2811 Fax: +81-3-3406-7345
E-mail: sales@hq.unu.edu   http://www.unu.edu